Studies in Computational Intelligence

Volume 1252

The series "Studies in Computational Intelligence" (SCI) publishes new developments and advances in the various areas of computational intelligence—quickly and with a high quality. The intent is to cover the theory, applications, and design methods of computational intelligence, as embedded in the fields of engineering, computer science, physics and life sciences, as well as the methodologies behind them. The series contains monographs, lecture notes and edited volumes in computational intelligence spanning the areas of neural networks, connectionist systems, genetic algorithms, evolutionary computation, artificial intelligence, cellular automata, self-organizing systems, soft computing, fuzzy systems, and hybrid intelligent systems. Of particular value to both the contributors and the readership are the short publication timeframe and the world-wide distribution, which enable both wide and rapid dissemination of research output.

Indexed by SCOPUS, DBLP, WTI AG (Switzerland), zbMATH, SCImago.

All books published in the series are submitted for consideration in Web of Science.

Roger Lee · Jongbae Kim · SungTaek Lee
Editors

IT Service and Business Innovation

Volume 1

Springer

Editors
Roger Lee
Software Engineering and Information Technology Institute
Central Michigan University
Mt. Pleasant, MI, USA

Jongbae Kim
Soongsil University
Seoul, Korea (Republic of)

SungTaek Lee
Yongin University
Seoul, Korea (Republic of)

ISSN 1860-949X ISSN 1860-9503 (electronic)
Studies in Computational Intelligence
ISBN 978-3-032-14396-9 ISBN 978-3-032-14397-6 (eBook)
https://doi.org/10.1007/978-3-032-14397-6

This Springer imprint is published by the registered company Springer Nature Switzerland AG
The registered company address is: Gewerbestrasse 11, 6330 Cham, Switzerland

Contents

Public Acceptance of Drone Delivery Services from Convenience Stores: A Case Study in South Korea

Woosik Jeong, Kunwoo Lee, KwangSup Shin, and Woojung Kim

1 Introduction

Drones were initially adopted mostly for military applications, but their usage has now expanded to a wide range of industries including agriculture, surveillance, and transportation and logistics among others. In particular, drone delivery services are leading the innovation of last-mile logistics by combining core technologies of the Fourth Industrial Revolution, and it can contribute to solving spatial imbalance issues in South Korea, such as food deserts and areas with limited road accessibility. Commercialization of drone deliveries is rapidly progressing both domestically and internationally, and the government is expanding policy support, such as the "K-Drone Delivery" initiative [1]. Drones are gaining attention as a fast and efficient delivery method in areas with limited accessibility. Furthermore, delivery drones are viewed as a sustainable mode to reduce negative externalities (e.g., emissions and traffic congestion) generated by conventional delivery vehicles [2].

Despite their clear advantages to our society, drones face an obvious technical constraint, i.e., their limited carrying capacity, which restricts them to transporting limited product types. Considering drones' technical limitation, they are particularly ideal for transporting small-sized and lightweight products. Thus, the integration of drone deliveries into convenience store industries—where products are typically small and light—would offer a promising and realistic application scenario.

W. Jeong · K. Lee · K. Shin · W. Kim (✉)
Graduate School of Logistics, Incheon National University, Incheon, Republic of Korea
e-mail: wkim@inu.ac.kr

W. Jeong
e-mail: jws1112@naver.com

K. Lee
e-mail: leegunwoo@korea.kr

K. Shin
e-mail: ksshin@inu.ac.kr

R. Lee et al. (eds.), *IT Service & Business Innovation*, Studies in Computational Intelligence 1252, https://doi.org/10.1007/978-3-032-14397-6_1

Given the growing importance of drone deliveries, the purpose of this study is to better understand consumer behaviors for drone deliveries (i.e., the willingness to use drone deliveries) using a discrete choice model. More specifically, this research focuses on convenience store users in South Korea, as products in convenience stores are mostly small-sized and lightweight, which makes them suitable for drone deliveries. The contributions of the present research are two-fold:

(1) While existing studies have primarily focused on general parcel deliveries, the present research examines drone deliveries from convenience stores, where products are typically small-sized and lightweight. These products are suitable for drone deliveries given the current drone capabilities.
(2) It provides data-driven evidence of consumer behaviors to support strategic planning for the future commercialization of delivery drones by using a discrete choice model.

The findings from the present research offer practical implications for identifying target service areas where drone delivery services are feasible as well as consumer segmentation groups with higher willingness to adopt the service. These insights can be a valuable input into strategic plans and tailored marketing strategies for drone delivery services from the perspective of convenience stores. This paper is organized as follows: the next section briefly provides background information on drone deliveries, and the third section reviews the existing literature on drone deliveries. The fourth section presents methodologies including survey data description and modeling technique. The fifth section presents results and discussion, and the last section concludes the present research.

2 Background

Drones (i.e., unmanned aerial vehicles) can be remotely controlled or even fly autonomously without a human pilot, and are used in various fields such as logistics, agriculture, and security [3]. International and domestic regulations classify drones based on their operational methods, weight, and intended use, though definitions may vary. However, all drones share the common characteristic of being unmanned aerial vehicles. Initially developed for military purposes, drones now have rapidly expanded into various industrial applications. Government agencies in various countries manage drones by categorizing them based on operational risk levels and functional capabilities.

Drones are classified based on various criteria such as takeoff weight and control methods, and are divided into fixed-wing, rotary-wing, and hybrid types according to their main shape. Rotary-wing drones have excellent maneuverability, while hybrid drones are highly versatile due to their energy efficiency and high-speed flight capabilities. With technological advances and lower barriers to entry, drones are rapidly spreading in both the private and public sectors. Governments around the world are nurturing drones as a core industry of the 4th Industrial Revolution.

The drone industry is rapidly expanding based on ICT convergence technology and is expected to grow to $87.5 billion by 2033. Originally developed for military use, drones are expanding their applications into civilian and industrial sectors, gaining attention as high-value-added platforms integrated with IoT, AI, and 5G. The industry is forming a structure that encompasses not only manufacturing but also operation, services, and data analysis in the downstream market. Drones are evaluated as a testbed for 4th Industrial Revolution technologies and a strategic core technology driving cross-industry convergence.

Countries around the world are utilizing drones to improve delivery efficiency and overcome geographical constraints. Malaysia, China, the United States, and Rwanda are actively introducing drones for medical, emergency response, and public services. Drones contribute to improving delivery speeds and addressing logistics blind spots, thereby recognizing their commercial and public value simultaneously. This trend demonstrates that drones are emerging as a core tool for global logistics innovation.

Domestic drone logistics services are expanding mainly in island and mountainous areas through cooperation between the government, the private sector, and local governments. The Ministry of Land, Infrastructure, and Transport is advancing the 'K-Drone Delivery' project to build infrastructure and operate demonstration cities. GS25, the Korea Post Corporation, and others are conducting pilot drone delivery services for essential goods and medications to verify technological stability and operational feasibility. This demonstrates that drones possess practical utility as a complementary means for last-mile logistics.

The commercialization of drone delivery requires not only technological advancements but also the establishment of legal and institutional frameworks, which have emerged as key challenges. The government is establishing operational standards for drone registration, pilot qualifications, and insurance through the Basic Plan for the Development of the Drone Industry and the Aviation Safety Act. Additionally, it is securing regulatory flexibility through systems such as drone special liberalization zones and demonstration cities. Standards for delivery service registration, regulations on high-risk flights, and personal information protection are also under development. These institutional frameworks serve as essential prerequisites for the nationwide expansion of drone delivery services.

3 Literature Review

The recent interest in drone delivery services has led to a growing body of research on consumer acceptance and intentions to use them. In particular, a number of studies have analyzed the perceptions, risk factors, and technology acceptance of drone delivery services, focusing on food delivery and general parcel delivery as summarized in Table 1.

Hwang, Choe (2019) explored the perceived risk factors of drone food delivery services and suggested factors to consider for successful service implementation [4]. Hwang, Lee, and Kim (2019) analyzed the impact of perceived innovativeness of

Table 1 A summary of the literature

Source	Product type	Geographical location	Influencing factors
[4]	Food (prepared meal)	South Korea	Perceived risks, image of drone delivery services
[5]	Food (prepared meal)	South Korea	Perceived innovativeness, attitude, gender, age
[6]	Parcel	Pakistan	Privacy issue
[7]	Parcel	Czech Republic	Expectancy, desire, attitude
[8]	Parcel	United States	Speed, risks, innovativeness

drone delivery services on consumer attitudes and behavioral intentions and empirically found that gender and age may have moderating effects on the relationship [5]. Khan, Tausif, and Javed Malik (2019) examined drone delivery acceptance and concern factors among urban consumers in Pakistan and reported that positive attitudes toward drone delivery technology were significantly associated with actual usage intentions [6]. Osakwe et al. (2022) analyzed the key factors that influence the intention to adopt last-mile delivery using drones and found that perception of delivery risk is an important variable in consumers' acceptance attitude [7]. Yoo, Yu, and Jung (2018) found that privacy concerns have a negative impact on the intention to adopt drone delivery services, while personal innovativeness has a positive impact [8]. They also emphasized that the determinants of acceptance may vary depending on the customer's living environment. These previous studies show that not only technological advancements, but also consumer emotional factors, social concerns, and local characteristics are the main variables that affect acceptance of drone delivery services.

While previous studies have focused on drone delivery services mostly for food and general parcel deliveries, this paper empirically analyzes the acceptance of convenience store-based drone delivery services considering the realistic conditions of delivering small and lightweight goods. A discrete choice model was used to quantitatively identify specific factors affecting respondents' willingness to use drones, and qualitative analysis was conducted through free-response questions. The feasibility-oriented research design, which reflects the weight limitations of drones, is different from previous studies. In particular, spatial scalability was also considered by assuming delivery bases in islands and mountainous convenience stores.

4 Methodology

4.1 Survey Description

This study conducted a survey targeting 300 adults in South Korea to examine the willingness to use drone deliveries, focusing on convenience store users. The questions include basic demographic and socio-economic information and the willingness to adopt drone delivery services from convenience stores. In addition, questions related to their major concerns over drone deliveries and situations in which consumers are willing to use such services, e.g., emergency medical supplies, were included in the survey.

The distributions of key variables are presented in Table 2. The respondents reflected various demographic characteristics, including gender, age, region, education level, occupation, and income. Over 60% of respondents were aged 40 or older, with a high proportion residing in Seoul and Gyeonggi Province, and the majority were employed in office-related occupations. The survey questions were designed to comprehensively measure attitudes toward drone delivery services, including usage intention (y_i; in the 5-level Likert scale from 'very unwilling' to 'very willing'), interest in technology, and price acceptance.

4.2 Discrete Choice Modeling

The discrete choice model is a technique for quantitatively analyzing individual choice behavior, which estimates the probability of choice based on the characteristics of the chooser and the attributes of the alternatives. This study analyzed the choice behavior of users of drone delivery services using the ordered logit model. It accurately identified the effects of socioeconomic characteristics such as gender, age, and housing type on choices. This model can be used as practical foundational data for drone delivery commercialization strategies and demand forecasting. The model can be expressed as presented in Eqs. 1 and 2 [9]:

$$U_i^* = X_i\beta + \varepsilon_i \tag{1}$$

where,

$\boldsymbol{U_i^*}$: Utility of respondent $\boldsymbol{i}$

$\boldsymbol{x_i}$:: Independent variable vector (e.g., gender, age, etc.)

$\boldsymbol{\beta}$: Coefficient vector.

$\boldsymbol{\varepsilon_i}$: Error term

$$y_i = \begin{cases} 1 \ if \ U_i^* \leq \mu_1 \\ 2 \ if \ \mu_1 < U_i^* \leq \mu_2 \\ J \ if \ \mu_{J-1} < U_i^* \end{cases} \tag{2}$$

Table 2 Summary statistics

Category	Variable	%
Gender	Male	50
	Female	50
Age group	20–29	15.7
	30–39	17.7
	40–49	21.7
	50–59	23.3
	60+	21.7
Area	Seoul	27.0
	Gyeonggi province	27.7
	Incheon	9.3
Education level	High school degree or less	17.3
	Associate degree	16.7
	College degree	57.0
	Graduate school degree	9.0
Residential type	Single-person household	16.3
	Living with family/others	83.3
Occupation	Office worker	43.7
	Self-employed	10.7
	Homemaking	14.3
	Students	4.0
Income level	20 million KRW or less	22.3
	20—39 million KRW	30.0
	40—59 million KRW	24.7
	60—79 million KRW	13.7
	80—99 million KRW	5.7
	100 million KRW or more	3.7

5 Results

According to the survey results, the main items purchased at convenience stores are beverages, snacks, and ready-to-eat meals. There was also high demand for drone delivery services, particularly for nighttime emergency deliveries. Drone deliveries are perceived as useful in emergency situations, though concerns about safety and delivery costs were also raised. Respondents prioritize logistics quality and economic benefits, and these expectations and concerns must be reflected in service design. This analysis provides practical grounds for developing commercialization strategies and demand-tailored policies for drone delivery.

The results of the ordered logit model are presented in Table 3. It revealed that technical familiarity, undergraduate/graduate students, and dormitory residence status significantly influenced the intention to use drone delivery services. Interestingly, the results showed that the technological familiarity negatively affected the intention. That is, individuals with lower familiarity exhibited a higher willingness to use the service. This counterintuitive finding may indicate that those who are unfamiliar with drones are more optimistic about the service and its benefits, while those with greater familiarity may be more aware of drones' technical constraints, limitations, and safety concerns. Education level was also found to have a positive impact on the willingness. Those with higher education levels exhibited higher willingness to adopt the service. The variable age negatively affected the willingness. Typically, younger individuals are more likely to adopt new technologies than others. Regarding the residential type, individuals residing in dormitories were less likely to use drone services. This is partly due to constraints such as restricted delivery access points within dormitory facilities. These results highlight the importance of user-specific characteristics in designing drone delivery services. Targeted strategies considering the key determinants identified from the model are essential to enhancing user acceptance.

In addition to the discrete choice model, the present research qualitatively analyzed respondents' opinions towards drone delivery services. The opinions are manually classified as presented in Table 4. As a result, the responses were grouped into five categories: (a) safety and security concerns, (b) accessibility and convenience, (c) speed and time expectations, (d) cost and payment issues, and (e) others. The results revealed that among the total 242 responses, the highest proportion (29.8%) was "safety/security concerns," reflecting significant anxiety about drone crashes, loss, and theft. "Accessibility and convenience" (13.2%) reflected concerns about delivery methods and service procedures, while "speed expectations" (5.0%) and "cost burden" (2.5%) were also mentioned by some respondents. The remaining "other" category (49.5%) included vague perceptions or incomplete responses, making it difficult to draw specific conclusions. Overall, users' expectations and concerns are mixed, and it is necessary to balance these factors in service design.

6 Conclusion

This study quantitatively and qualitatively analyzed the factors influencing the intention to use drone delivery services among convenience store consumers. The present research analyzed the factors affecting the acceptance of drone delivery services using an ordinal logit model. As a result, technical familiarity, residential type, and occupational group had a significant impact on the intention to use the service, and college students showed high acceptance. Dormitory residents had a low intention to use the service due to delivery environment constraints. The qualitative analysis pointed out issues such as security, accessibility, and reliability, suggesting the need for user-customized design.

This study focused on analyzing the acceptability of drone delivery, but it had limitations such as sample bias and insufficient representation of the elderly and

Table 3 Final model results

Independent variable	Model		
	Coefficient	z-value	p-value (p > |z|)
Female	−0.351	−1.42	0.155
Age	−0.129	−1.34	0.179
Technical familiarity	−0.541	−4.42	0.000
Education level (Base: Highschool Graduated)			
College graduated	0.453	1.24	0.217
University graduated	0.289	0.97	0.332
Graduate school graduated	0.683	1.55	0.122
Job (Base: Office work)			
Skilled/labor work	−0.160	−0.40	0.687
Self-employed	−0.323	−0.86	0.392
Full-time homemaker	0.538	1.42	0.156
Student	1.325	2.14	0.032
Unemployed	0.187	0.45	0.652
Other	0.637	1.57	0.117
Living Environment (Base: Single)			
Living with family	−0.358	−1.15	0.249
Living in a dormitory	−3.772	−2.25	0.024
Model Parameters			
$\mu(1)$	−4.778		
$\mu(2)$	−3.241		
$\mu(3)$	−1.859		
$\mu(4)$	0.581		
Number of observations	300		
Log likelihood	−406.23		
LR chi^2 (14)	50.11		
Pseudo R^2	0.058		

Table 4 Opinions toward the drone delivery service

Category	Number of responses	Percentage of the respondents (%)
Safety/security concerns	72	29.8
Accessibility/convenience	32	13.2
Speed/time expectations	12	5.0
Cost/payment issues	6	2.5

rural areas. Although the intention to use was measured, there was a lack of analysis of actual choice behavior, and the study also failed to reflect supplier-side factors. Qualitative opinion analysis also faced interpretation limitations due to constraints in response expressions. Despite these limitations, this study holds significance for its practical contributions to initial demand assessment and the identification of user factors.

Acknowledgements This work is supported by the Korea Agency for Infrastructure Technology Advancement (KAIA) grant funded by the Ministry of Land, Infrastructure and Transport (Grant RS-2022-00156324). An earlier version of this research was presented at the 3rd ACIS International Symposium on Emotional Artificial Intelligence & Metaverse on June 26th, 2025 in Busan, South Korea. This work is based on the first author's master thesis submitted to Incheon National University.

References

1. MOLIT (2024) Delivery service by K-Drones launches in Island Areas from August
2. Goodchild A, Toy J (2018) Delivery by drone: An evaluation of unmanned aerial vehicle technology in reducing CO_2 emissions in the delivery service industry. Transp Res D Transp Environ 61:58–67. https://doi.org/10.1016/j.trd.2017.02.017
3. Garg V, Niranjan S, Prybutok V, Pohlen T, Gligor D (2023) Drones in last-mile delivery: a systematic review on efficiency, accessibility, and sustainability. Transp Res D Transp Environ 123:103831. https://doi.org/10.1016/j.trd.2023.103831
4. Hwang J, Choe JY (2019) Exploring perceived risk in building successful drone food delivery services. Int J Contemp Hospit Manag 31(8):3249–3269. https://doi.org/10.1108/IJCHM-07-2018-0558
5. Hwang J, Lee JS, Kim H (2019) Perceived innovativeness of drone food delivery services and its impacts on attitude and behavioral intentions: the moderating role of gender and age. Int J Hosp Manag 81:94–103. https://doi.org/10.1016/j.ijhm.2019.03.002
6. Khan R, Tausif S, Javed Malik A (2019) Consumer acceptance of delivery drones in urban areas. Int J Consum Stud 43(1):87–101. https://doi.org/10.1111/ijcs.12487
7. Osakwe CN, Hudik M, Říha D, Stros M, Ramayah T (2022) Critical factors characterizing consumers' intentions to use drones for last-mile delivery: does delivery risk matter? J Retail Consum Serv 65. https://doi.org/10.1016/j.jretconser.2021.102865
8. Yoo W, Yu E, Jung J (2018) Drone delivery: factors affecting the public's attitude and intention to adopt. Telematics Inform 35(6):1687–1700. https://doi.org/10.1016/j.tele.2018.04.014
9. Ortúzar JDD, Willumsen LG (2011) Modelling transport, 4th edn. Wiley, Hoboken, NJ

A Study of Data Verification Processes and Frameworks to Ensure Reliability of AI Results

Dongju Ryu, SeungHee Kim, and Sungtaek Lee

1 Introduction

Artificial intelligence (AI) has emerged as a fundamental technology shaping both current and future innovations [1]. Particularly, the development of generative AI has highlighted the increasing demand for high-quality training datasets. AI applications are being rapidly integrated and expanded across a wide range of industries, including security, video analysis, speech-to-text conversion (natural language processing), autonomous driving, media, transportation, environmental monitoring, healthcare, manufacturing, financial services, and entertainment. This growth has concurrently propelled the development of the AI training data industry [2].

AI training data encompasses not only structured datasets but also unstructured ones, which are collected and labeled through annotation processes that incorporate Ground Truth (GT) [3]. This data serves as the foundation for supervised and unsupervised learning algorithms. However, the emergence of unsupervised learning methods, such as those utilized in ChatGPT, has brought to light critical issues related to the origin and reliability of datasets. The accuracy and trustworthiness of AI outputs are heavily dependent on the quality of the training data employed, underscoring the importance of ensuring precise and reliable data for AI development.

D. Ryu (✉)
Department of Convergence Security Engineering, Sungshin Women's University, Seoul-City, Republic of Korea
e-mail: ryu@btress.com

S. Kim
AI Planning Division (AI Lab.), Go-To Guy Company, Gyeonggi-Do, Republic of Korea
e-mail: zelssen@naver.com

S. Lee
Department of Business Administration, Yong In University, Gyeonggi-Do, Republic of Korea
e-mail: totona22@yiu.ac.kr

R. Lee et al. (eds.), *IT Service & Business Innovation*, Studies in Computational Intelligence 1252, https://doi.org/10.1007/978-3-032-14397-6_2

This paper introduces the concept of "AI Data Verification," defined as the process of securing legally compliant and appropriately vetted training data for AI. This verification process involves examining and validating datasets across all stages of AI development to produce valuable and dependable results. By focusing on purpose-driven development and combining accurate data with advanced AI models, trustworthy outcomes can be achieved.

This paper proposes a data verification process for AI training, focusing on validating and re-evaluating data throughout all stages, including collection, normalization, labeling, and model training [4]. The study applies this proposed framework to real-world implementation processes, emphasizing the use of unstructured data over structured data in AI system development. Depending on the purpose of AI usage, data is first collected and normalized. Subsequently, the data undergoes a labeling process to make it comprehensible to AI systems. Once labeled, the data is trained on a pre-selected AI model, completing a single development process.

Throughout this process, the study identifies the stages and actions necessary to minimize errors in data collection, normalization, labeling, and model outcomes, ensuring the highest quality of training data. The proposed framework has shown to play a vital role in improving the reliability and accuracy of AI learning results while reducing potential legal challenges associated with training data.

Chapter 1 provides an overview of the study's purpose and research scope. Chapter 2 examines the characteristics of AI data and specifies the requirements for targeted datasets. Chapter 3 introduces step-by-step data verification methods along with experimental results from their application. Finally, Chap. 4 concludes with suggestions for future research and practical applications.

2 Prior Studies and Their Limitations

2.1 Previous Literature

According to the suggestions provided by OpenAI, various features are introduced to enhance data learning capabilities by addressing challenges such as hallucinations, biases, and the lack of interpretability. These features are referred to as RLHF (Reinforcement Learning from Human Feedback), and the detailed verification process and procedures are illustrated in Fig. 1 [5].

- Supervised Fine-Tuning (SFT)—To train the model with human-intended poli cies, a small dataset curated by human labelers (or trainers) is used to fine-tune a pre-trained language model (LM).
- Reward Model (Mimic Human Preferences)—Human labelers collect a dataset by ranking and scoring multiple answer candidates generated by the SFT model in step 1, based on which answers are better. This dataset is used to train a new Reward Model.

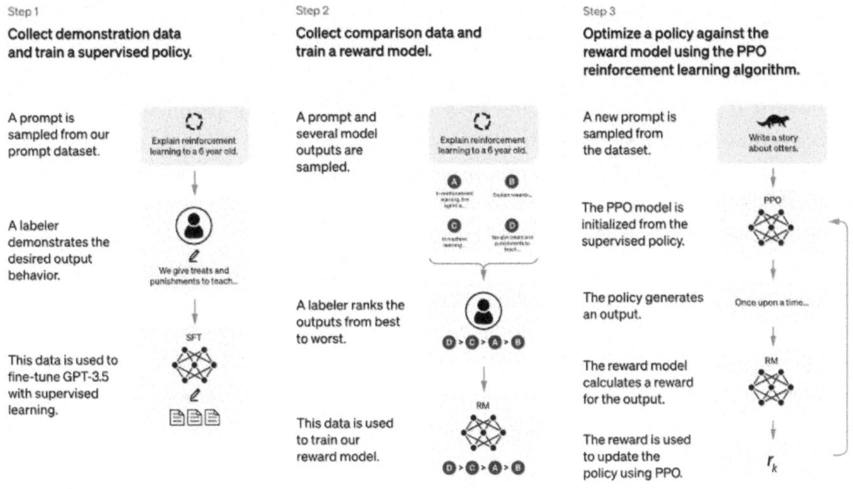

Fig. 1 RLHF applied to ChatGPT in OpenAI [5]

- Reinforcement Learning with SFT Model Using Proximal Policy Optimization (PPO)—The SFT model is exposed to various user inputs and undergoes reinforcement learning while interacting with the Reward Model.

OpenAI's RLHF (Reinforcement Learning with Human Feedback) technique is utilized to prevent data redundancy, minimize hallucinations, and correct unintended errors in data. It employs reinforcement learning with reward mechanisms while minimizing human intervention. Furthermore, addressing the drawbacks related to errors or biases stemming from subjective human judgments is necessary for its improvement [5].

2.2 *Limitations of Previous Study*

The methods for data validation identified through prior research are most commonly utilized by OpenAI. However, this approach does not provide a clear explanation or validation process for the collection and refinement of data. Additionally, most AI systems have concentrated research and investment primarily on model-centric reliability and robustness, focusing on in-depth algorithms. Based on this, they emphasize the significance of their research using solutions derived from mathematical formulas and visualizations through graphs.

Furthermore, recently, XAI (Explainable AI) has emerged as a method for understanding the data learning process and resolving internal data issues during the learning process [6]. However, this research faces numerous technical limitations, and it is believed that a comprehensive understanding of the learning and comprehension processes within models is essential to achieve reliability and robustness in practical AI results. However, while I acknowledge the importance of such research, I believe it is crucial to understand how data is collected in practice and how the collected data undergoes normalization processes to include valuable information.

Thus, I aimed to ensure the reliability of data by presenting specific procedures and analysis results to correct or resolve errors in data collection, normalization, and labeling before the learning phase begins.

Furthermore, the problems and solutions identified in these steps are expected to provide clear answers regarding the outcomes of AI learning.

Ultimately, data verification involves filtering out false information or personal data during the data collection and normalization process. If such processes are repeated mechanically, with both humans and machines continuously performing them, I believe more accurate data can be generated.

3 Data Validation Process and Framework

The experiment period of this study was from 2020 to 2023, and the types of data used in the experiment totaled 413. The total volume of the validation data was approximately 120 terabytes. The majority of the data consisted of unstructured formats, including videos, images, healthcare images and data, text, voice, natural language, sound, 3D data, generative data, sensor data, graphs, signals, digital waveforms, behavioral analysis data, and anomaly analysis data [7].

To enable AI to learn effectively from such a diverse range of data, it is crucial to thoroughly analyze the processes undertaken prior to each training session and identify the necessary datasets. Erroneous, inappropriate, or inaccurately provided data may potentially be used as noise data; however, this should only be done after verifying the accuracy and reliability of the data.

3.1 Suitability of the Model's Results

The data verification process proposed in this paper has already been adopted by many companies and organizations. As a result [8], it has become clear that verifying the data output at each stage of the AI implementation process is necessary [9].

The proposed data verification process does not introduce new technology or suggest new stages. Instead, it combines agile and waterfall methodologies, which are already used in software engineering, to create an integrated process.

Figure 2 illustrates the roles and key aspects that need to be examined at each stage of the proposed data verification process. However, given the complexity and vast amount of information involved, the figure may not be able to fully capture

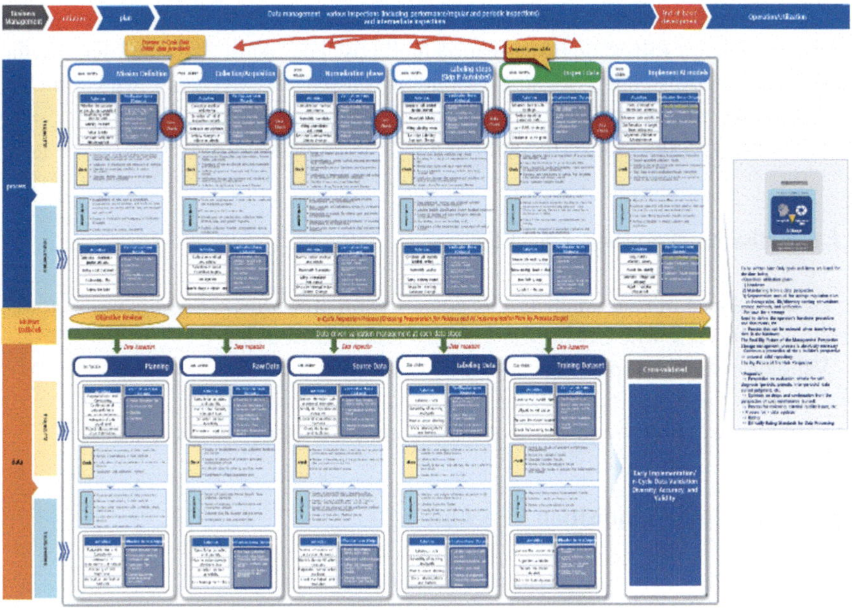

Fig. 2 Proposed data verification process and task-specific framework

everything accurately. Therefore, the key actions required at each stage are briefly outlined for clarity.

The process is structured to anticipate the expected data outcomes at each stage and to verify whether they align with requirements and objectives by using predefined checklists and reviewing the generated data [10, 11].

Each stage utilizes a pre-established checklist. This checklist is designed in advance to align with the purpose of the data, ensuring that verification is conducted objectively and quantitatively rather than subjectively. As a result, the predefined checklist is of utmost importance and must be presented in a well-structured document, backed by extensive experience and knowledge. The proposed data verification process and framework consist of six main stages. The verification flow is designed as a cyclical structure centered around the verification data. At each stage, the verification is conducted using checklists, and the iterative deliverables are passed on to the next stage.

Particularly, the process evaluates the consistency of identified issues and requirements within each stage numerically, ensuring a more objective scoring system and content. The upper section presents the roles and checklists of the process, while the lower section is structured to focus on reviewing data. This separation between the process and the data aims to clearly delineate roles and responsibilities. In the content of Fig. 2, the red arrow at the top represents the process known as the "n-cycle.“ This refers to a cyclical inspection method used during data verification. Before the training data is fully established, it illustrates the process of periodically verifying and providing feedback to ensure the data aligns with its intended purpose, using either sampling or full-scale investigation methods. The "n" in "n-cycle" denotes the number of iterations, and typically, conducting one to three cycles of inspection has been analyzed to yield reliable data.

A summary of the key verification items for each stage is as follows.

3.1.1 Mission Definition Checklist

- Establishment of roles such as data construction procedures, organizations, tools, etc., and review whether they are managed and carried out
- Validation of developers and adequacy of methods
- Checklist for matching checklists in various documents
- Checking whether the purpose of the project execution plan is consistent.

3.1.2 Collection/Acquisition Checklist

- Review of raw data collection methods and standards
- Legal System Review (Personal Information, Portrait Rights, Copyright)
- Prevention of bias in collected data, data standards and indicators
- Collection/Inspection Standards and Organization Review
- Verification through the existence and matching of various outputs (currentization)
- Collection Tools/Storage Environment Review
- Systematic establishment of data collection methods and standards, progress
- Consistency in the business plan
- Identification of collected data collection items, diversity, bias, and security measures
- Realistic collection Whether environmental data is synchronized.

3.1.3 Refinement/Normalization Checklist

- Review of original data purification methods and standards
- De-identification criteria such as personal information, result confirmation
- Refinement/Inspection Standards and Organization Review
- Verification of implementation companies and review of the adequacy of verification measures
- Checklist in each document (checklist)
- Collection Tools/Storage Environment Review
- Data refinement method and standard systemic establishment and implementation
- Data cleansing and redundancy review by comparing information
- Preparation of manuals for refining work and review of management
- Inspection and feedback on the inclusion of essential information in the source data

- Request data review of purification data and perform feedback

3.1.4 Labeling Checklist

- Review data labeling methods and criteria
- Bounding box, key point, segmentation, transcription, etc.)
- Review data types and input value ranges
- Accuracy (semantic accuracy, syntactic accuracy)-centered test
- Verification through the existence and matching of various outputs (currentization)
- Labeling Tools/Storage Environment Review
- Data refinement method and standard systemic establishment and implementation
- Labeling English Classification System Standard Inspection
- Review of labeling self-data verification methods (consistency, accuracy)
- Use labeling tools and how they work
- Verification of the existence and consistency of various outputs

3.1.5 Inspect Data Checklist

- Check whether there is an inspection of source data and labeling data
- Check the distribution of nonconformity data
- Data inspection results and quantity check according to data inspection standards
- Existence and consistency of various final (including intermediate and stage) outputs
- Data Validation Interim Results
- Inspection of original data and labeling data
- Partial optimization inspection by step to check the distribution of nonconformity judgment data (through checking the error rate and characteristic distribution of data)
- Check if the requirements provide datasets for training
- Uniformity inspection of inspection indicators and standards by third-party institutions

3.1.6 Implimentation AI Models Checklist

- Algorithmic Performance Measurement Inspection
- Review algorithm validation results
- Checking the results of n-cycle verification (request for expert advice if necessary)
- Final Data Model Application Results Inspection
- Verification of comparison of results compared to goals in the project execution plan
- Algorithmic Performance Measurement Inspection
- Dataset data error and issue content analysis that can be used for model and service development

- Final Data Model Application Results Inspection
- Review of flexibility in model selection and application

This paper verifies data based on predefined step-by-step rules and validation criteria. Experts with extensive experience in implemeting and operating various artificial intelligence systems participated in the process. They collaboratively reviewed whether the data and documents stored during each stage of the actual process were clearly presented as objective evidence and accurately represented.

In the next chapter, the proposed data validation process and framework will be used to analyze the data and present the results. Due to the large number of analyses and outcomes, only a few simple examples will be provided for illustration.

4 Results of the Proposed Data Validation

4.1 AI Data Conformity Check Results

In this chapter, we performed a root cause analysis of issues related to the appropriateness and accuracy of the data using the previously proposed data validation process. Furthermore, we identified methods to improve already implemented data and pinpointed problems at each stage to propose measures to prevent future issues. Through this, it was demonstrated that carrying out the data validation process ensures reliability in both the initial construction and future operation of the data.

Table 1 briefly summarizes portions of the data used in the experiments, highlighting the need for a data verification process emphasized in this study. Table 1 provides a portion of the overall data used in this study, presented as a table summarizing the evaluation scores from the defined checklist and the results of examining the data at each stage. Regardless of the type and quantity of the data, it reflects the outcomes of performing n-cycles and visually verifying the actual data. For validation conclusions, to account for potential biases, the verification results reflect the collective opinions of three or more individuals. Results for each item are marked with "X", "○", or "△", listing only the identified problematic parts. This represents actual examples of the results of listing and performing quantitative evaluations.

4.1.1 Proof with Insufficient Data

This paragraph presents evidence verified through a data validation process aimed at enhancing the reliability of AI outputs.

Figure 3 shows that 69 types of AI datasets have already been constructed. However, it was concluded that the training data and the results were highly unsatisfactory. Since the root cause was unclear, an analysis of datasets and documents for each process was conducted.

Table 1 AI data conformity check results (summary)

category		Match requirements (Purposefu		Data collection steps				Normalization steps					Labeling steps			Validation adequacy
Data Type	Data Name (omitted)	1-1	1-2	1-3	1-4	1-5	1-6	1-7	1-8	1-9	1-10	1-11	1-12	1-13	1-14	1-15
Vision (3D+Video)	3D human interaction data (2 and 3 or more)		△										△			
	AI-based drawing data construction for children's art												X			
Healthcare (Image)	Images and Clinical Data		△		△								△			
Vision + Sensor	Field crops (cabbage, etc.) cultivation data		X		X								X			△
	Rice growth anomaly recognition data		X		X								X			△
	Ocean current seed production (seaweed) data		X		X		X				X	X	△			△
	Shellfish seed production (abalone) data		△		X											△
	Intelligent insect smart farm (silkworm, double cricket) data		X		X											△
	Traditional Food Fermentation Convergence Data		X		X								X			△
	Dog, cat health information data		X		X	X	X				X	X				
	Cow (Korean beef, dairy cow) and pig estrus behavior data		X						△		△		X			△
	Citrus Sweep Learning Data		X		X				X				X			X
Safety + Image	Face parsing Data		△		△						△		△			△
	Facial Recognition Aging Image Data		△		△						△					△
	IR real-world facial detection and recognition data		△		△						△					△
	Furniture, home appliances, and office equipment use motion capture data		X				△	△	△	△	X	X	X			△
	Conversation/Presentation Motion Capture Data	△	X				△	△	△	△	X	X	X			△
	Multi-image identical situation and object identification data		△		X	△	X				△	△	△			△
	X-ray Multi-Object Recognition Data		△			△							△			
	Exploration data for underground storage of greenhouse gases	△									△		△	△		
	Solar power generation status and site analysis data		X										X			X
	Water source-intake source integrated water quality and algae data		X								△		X			
	3D Ground Penetrating Radar Exploration Data		△		△				△		△		X			
	Building Wind Disaster Risk Analysis Data				△						△		X			
	Construction Sand Quality Control for Building Collapse Disaster Prevention		△						△	△			△			
3D+Sensor+Text+Image	Digital K-Art Data				△	△	△					△	△	△		
	Exhibition performance docent data				△	△	△					△	△	△		
Need to supplement (△)		14		16				22					13			13
Fell short (X)		12		13				9					12			2
sum		26		29				31					25			15

The analysis revealed errors in the labeling data for 60 out of the 69 types of datasets. Furthermore, the data cleansing (normalization) process failed to adhere to proper expressions and rules, which significantly impacted AI learning.

According to the overall evaluation, it was confirmed that proper data validation had not been conducted, and problems were ignored during training, leading to unrealistic AI results. As illustrated in Fig. 4, additional analyses were performed to identify and address the issues.

Although the implementing company was expected to reflect upon the feedback and make improvements, as shown in Fig. 5, six months later, the identified issues remained unresolved.

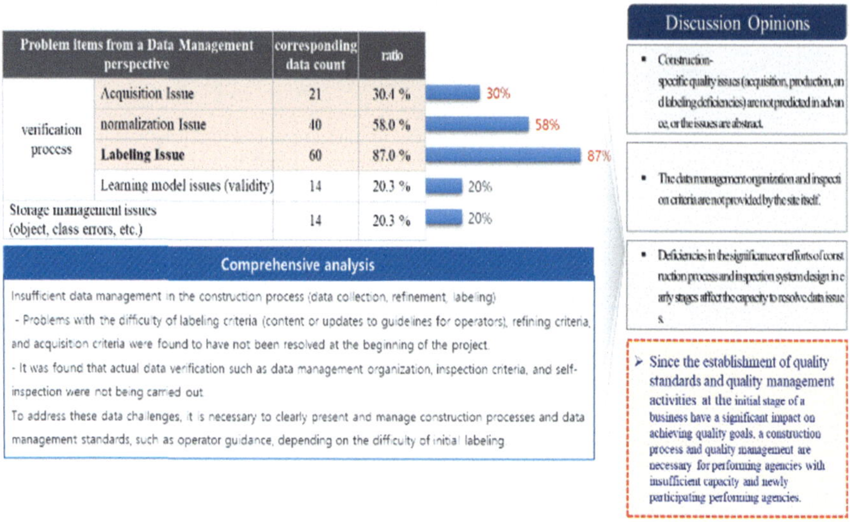

Fig. 3 Verification and analysis results for 69 types of data

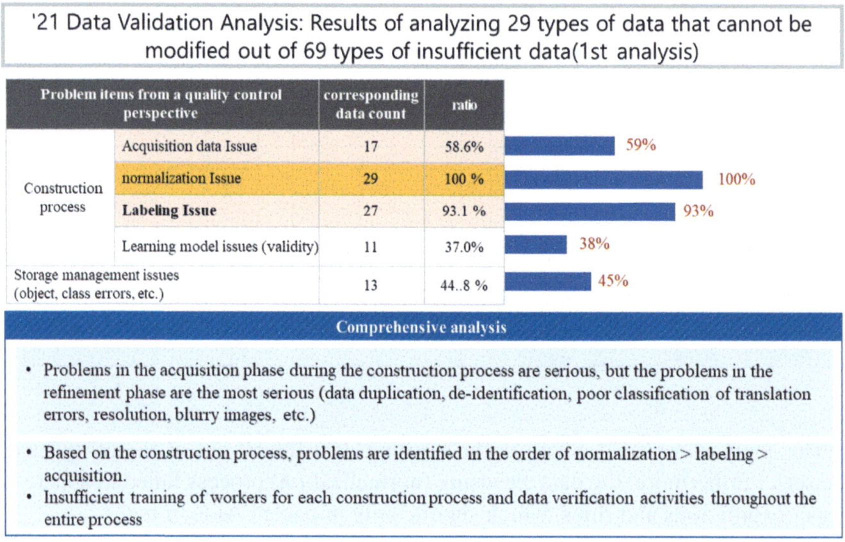

Fig. 4 Step-by-step analysis results based on data verification errors in the first experiment

The main reasons for this include the failure to initially organize and store data appropriately for step-by-step training, resulting in missed opportunities, and the existence of datasets affected by seasonal and time-related factors, which required waiting an additional year for data collection. Additionally, some datasets remained

'21 Data Validation Analysis: Results of analyzing 12 types of data that cannot be modified out of 69 types of insufficient data (2nd analysis)

Problem items from a data control perspective		corresponding data count	ratio
Construction process	Acquisition Quality Issue	6	50.0 %
	normalizationQuality Issue	12	100.0 %
	Labeling Quality Issue	12	100.0 %
	Learning model quality issues (validity)	6	50.0 %
Storage management issues (object, class errors, etc.)		6	50.0 %

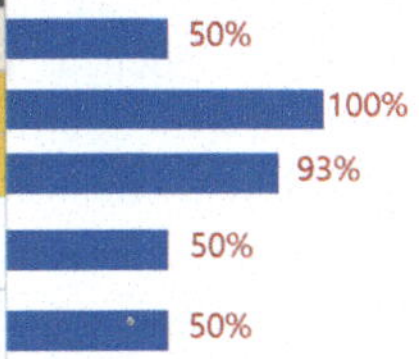

Comprehensive analysis

- Insufficient data inspection and management by process of implementation agency construction
- Insufficient response to problems of labeling data such as purification and semantic accuracy
- Failure to meet target indicator criteria due to insufficient capacity in verification through validity
- The result is that the focus is on project completion centered on outputs based on the unilateral data judgment of the implementation agency due to insufficient basis for objective judgment on semantic accuracy

Fig. 5 Secondary data verification and analysis results

unresolved due to these factors, and some companies abandoned efforts to reconstruct new datasets due to the recognition of high costs involved.

Ultimately, through thorough validation and experimentation, it was concluded that achieving reliable AI outputs necessitates the proper implementation and step-by-step training of datasets.

5 Conclusion

This study proposes a process and framework for validating the data used in AI model training to ensure the reliability of AI-generated results. The proposed process and framework identify potential data issues at each stage of the data construction process and suggest methods to ensure the accuracy and reliability of AI models. Additionally, step-by-step data improvement measures are presented to address these issues.

The study analyzed the construction procedures and components of the data used for training AI models, highlighting the importance of each stage and emphasizing the need for continuous research in this area. Data error types identified include file errors, image errors, and compression errors. Considering that high-quality data accounts for approximately 80% of the overall performance of AI training datasets, this study implemented systematic data management tailored to the requirements of each construction stage. As a result, issues such as errors and problems with the sources of collected data were resolved, and high-quality data was successfully secured. To enhance the reliability of AI training results, it is essential to provide accurate information that adheres to normalization standards and aligns with the goals of AI models from the initial data collection stage. It was found that collected data with a clear purpose tend to be more accurate and hold greater value as data.

In particular, from the design stage of the data construction process, clear goal setting and rigorous data validation methods must be established. This includes sequential stages such as mission definition, data collection, data normalization, data labeling, and model training. Maintaining consistency in data formatting is also crucial, as different construction methods or tools within the same data category can result in varying file formats.

This study proposes effective validation measures to identify potential data quality issues at each stage of the data validation process and presents a comprehensive approach to ensuring reliability. By applying the proposed process and framework, the study is expected to contribute to significantly improving the reliability and accuracy of AI model outputs while minimizing biases and hallucinations.

References

1. Yoo S, Cho S, Song M, Jeon S, Lim S, Jung S, Park S, Park G, Kim H, Lee D (2018) A study on improved label recognition method using deep learning. In: Proceedings of Spring Academic Conference. Seoul National University of Science and Technology, TechWing Inc.
2. National Information Society Agency (NIA) (2021) AI training data construction project management manual
3. Kim S, Ryu D (2023) A method for enhancing the quality of video/image data for AI learning of unstructured data. J Converg Secur 23(2):1–10
4. Korea Internet & Security Agency (KISA), Status of Intelligent CCTV Certification. https://www.kisa.or.kr
5. https://openai.com
6. https://www.anthropic.com/research/mapping-mind-language-model
7. Ministry of Science and ICT NIA, and TTA, Guidelines for AI Training Data Quality Management, ver. 3.0 (2023)
8. Ministry of Science and ICT, NIA, and TTA, AI Training Data Quality Management Guidebook, vol. 1: Quality Management Construction Manual (2021)
9. Ministry of Science and ICT, NIA, and TTA (2022) Guidelines for AI Training Data Quality Management, ver. 2.0, vol. 2: Data Construction Manual, p 1
10. Ministry of Science and ICT and NIA (2022) Announcement on the 2022 AI Training Data Construction Support Project
11. Chang-Yeop L, Hong-Jae L, Yong-Rak C (2018) Establishing a data quality management system focused on disease data. J Korea IT Policy Manag Soc

Discriminant Analysis of Career Types According to Personality Types of Cyber University Students: Focused on GOLDEN Personality Test and Holland's RIASEC Model

Jungjae Lee and Youkyung Lee

1 Introduction

Career choices significantly affect individual life satisfaction, self-realization, and occupational sustainability. With the expansion of digital learning environments, there is increasing demand among adult learners seeking to balance education and career paths beyond traditional university settings. Cyber university students represent a diverse and dynamic group with varied motivations such as self-directed career planning, career transitions, and recovery from career disruptions. However, research on personality-based career exploration for this group remains insufficient. Existing studies primarily focus on correlations between MBTI personality types and Holland career types, predominantly involving high school and traditional university students. Empirical studies employing the GOLDEN Personality Type Profiler and focusing on cyber university students are rare. Furthermore, statistical attempts to classify or predict career types based on personality sub-dimensions are nearly nonexistent. This study aims to address this gap by empirically analyzing the structural relationship between personality types and career types among cyber university students using discriminant analysis.

The purpose of this study was to identify the impact of personality types on career type predictions among cyber university students, thereby providing foundational data for career counseling and educational support. Specific objectives were: To assess the distribution of GOLDEN personality dimensions among cyber university

J. Lee
Department of Entertainment, Soongsil Cyber University, Seoul, Republic of Korea

Y. Lee (✉)
Department of Christian Counseling Walfare, Soongsil Cyber University, Seoul, Republic of Korea
e-mail: hontas!@mail.kcu.ac

R. Lee et al. (eds.), *IT Service & Business Innovation*, Studies in Computational Intelligence 1252, https://doi.org/10.1007/978-3-032-14397-6_3

students. To explore the relationship between GOLDEN personality dimensions and Holland career types (RIASEC). To statistically validate the ability of personality dimensions to significantly classify career types through discriminant analysis.

< Research Questions >

What is the distribution of GOLDEN personality dimensions among cyber univer sity students?

How are GOLDEN personality dimensions related to Holland career types (RIASEC)?

2 Theoretical Background

2.1 GOLDEN Personality Type Profiler

Understanding the relationship between career types and personality types requires an integrated theoretical framework that encompasses structural dimensions of personality and occupational classification systems. This study adopted the GOLDEN Personality Type Profiler and Holland's RIASEC model as its main theoretical frameworks to analyze whether career types could be differentiated according to personality dimensions.

First, the GOLDEN Personality Type Profiler is based on Jung's psychological types theory, structurally measuring the four dimensions utilized by the MBTI (Energy Direction, Information Gathering, Decision Making, and Lifestyle) to classify individuals into 16 personality types. The GOLDEN assessment has been widely applied in practical domains such as job suitability, organizational communication, and career counseling, particularly valued for predicting occupational behaviors and performance [1] 2.

Each personality dimension connects with career-related psychological characteristics as follows: The Energy Direction dimension suggests that Extraversion is suited for occupations involving significant social interaction (e.g., E/S types), whereas Introversion is aligned with analytical and independent work (e.g., I/R types). The Information Gathering dimension indicates that Sensing types prefer concrete procedures and precision required in Realistic (R) or Conventional (C) occupations, while Intuition types are inclined towards Artistic (A) or Enterprising (E) occupations that emphasize creativity and planning. Thinking types prefer analytical, logical problem-solving linked to technical or managerial roles, whereas Feeling types align with value-oriented, interpersonal roles (S/A types). Lastly, Judging types thrive in structured, plan-oriented environments (C/E types), while Perceiving types excel in flexible and unpredictable contexts (A/I types).

2.2 Holland's RIASEC Model

Holland's RIASEC model categorizes vocational interests into six types: Realistic, Investigative, Artistic, Social, Enterprising, and Conventional [3]. Each type corresponds to particular personality traits and behavioral tendencies. Realistic (R) types prefer practical and physical activities, Investigative (I) types favor analytical and autonomous roles, Artistic (A) types thrive in emotional and unstructured settings, Social (S) types excel in cooperative, relationship-focused roles, Enterprising (E) types are competitive and initiative-driven, and Conventional (C) types prefer organized, rule-oriented environments. RIASEC theory is centered around the concept of congruence between personality and occupation, aligning with Super's [4] career development theory and Dawis and Lofquist's [5] theory of person-job fit [6]. Super regarded career choices as an expression of self-concept, suggesting personal values and personality significantly impact occupational decisions. Dawis and Lofquist explained job stability and satisfaction through the congruence between individual needs and job characteristics. Both theories emphasize the importance of the interaction between internal psychological factors and external environments, underpinning Holland's congruence theory.

Numerous existing studies have explored the relationships between MBTI and RIASEC, reporting tendencies such as Sensing (S) types associating with Realistic (R) or Conventional (C) careers, and Feeling (F) types correlating with Social (S) or Artistic (A) careers [7] 1. However, empirical research investigating discriminative relationships between GOLDEN dimensions and career types is sparse, particularly among cyber university students. This study employed discriminant analysis to explore these relationships. Discriminant analysis identifies combinations of independent variables that differentiate two or more groups and predicts the likelihood of a new case belonging to a particular group. It is particularly effective when dependent variables are clearly categorical and independent variables are continuous or binary. In this research, career types (RIASEC) were set as dependent variables, while the four personality dimensions served as independent variables to distinguish personality characteristics associated with each career type. Discriminant analysis allows verification of the model's classification power and explanatory strength using various statistical indicators such as Wilks' Lambda, discriminant function coefficients, and classification accuracy. It can also assess the predictive accuracy and the contribution of each independent variable, making it broadly applicable in career counseling and educational psychology fields.

3 Research Method

3.1 Participants

This study involved 200 adult learners enrolled at a cyber university during the summer and winter semesters of the 2024 academic year. The final analysis included 152 respondents who completed both the GOLDEN Personality Type Profiler and the Holland Career Aptitude Test.

3.2 Instruments

- GOLDEN Personality Type Profiler: Comprises four dimensions—Energy Di rection (Extraversion/Introversion), Information Gathering (Sensing/Intuition), Decision Making (Thinking/Feeling), and Lifestyle (Judging/Perceiving)—with each dimension scored through dichotomous choice items.
- Holland Career Aptitude Test: Based on the RIASEC model, participants' pri mary career types were determined by the highest scores among Realistic (R), Investigative (I), Artistic (A), Social (S), Enterprising (E), and Conventional (C).

3.3 Data Analysis Method

A multivariate Linear Discriminant Analysis (LDA) was conducted using the four dimensions of the GOLDEN Personality Type Profiler as independent variables and Holland's career types (RIASEC) as dependent variables. The analysis evaluated the model's classification accuracy, primary discriminant function coefficients, Wilks' Lambda, Eigenvalue, and Canonical Correlation. SPSS and Python-based statistical packages were used for data analysis.

3.4 Results

3.4.1 Overall Classification Accuracy and Model Summary

The overall classification accuracy of the discriminant model was confirmed to be 53.95%, indicating a higher-than-random predictive capability and suggesting that personality dimensions alone could reasonably classify career types (Table 1 and Fig. 1).

This graph illustrates the distribution of GOLDEN personality types among the 152 cyber university students participating in the study. It shows a high proportion of

Table 1 Summary of discriminant analysis results

Measure	Value
Eigenvalue (Function 1)	0.32
Canonical correlation	0.49
Wilks' lambda	0.84
Chi-square	24.67
df	20
p-value	0.081

The overall significance of the model is p = 0.081, indicating statistical significance at a near-significant level.

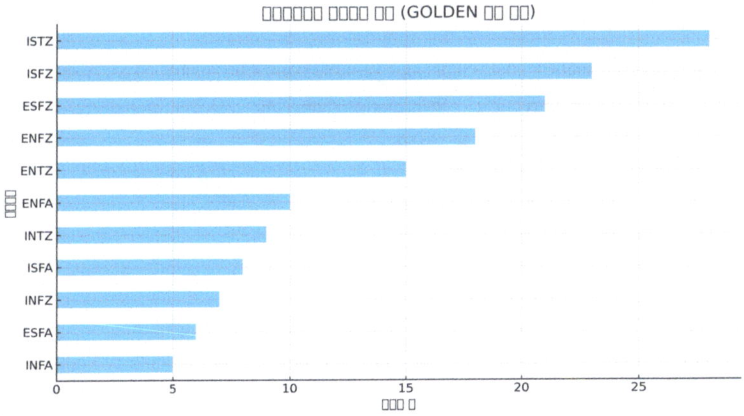

Fig. 1 Distribution of GOLDEN personality types

combined Judging (Z) and Feeling (F) types (e.g., ISTZ, ISFZ, ESFZ), which may closely relate to preferences for Conventional (C) career types.

3.4.2 Classification Accuracy by Career Type

See Table 2.

3.4.3 Discriminant Function Coefficients and Variable Contributions

See Table 3.

Table 2 Classification accuracy by career type

Career type code	Career type (RIASEC)	Number of samples	Correctly classified	Classification accuracy (%)
0	Realistic (R)	27	0	0.0
1	Investigative (I)	23	0	0.0
2	Artistic (A)	8	0	0.0
3	Social (S)	8	0	0.0
4	Enterprising (E)	4	0	0.0
5	Conventional (C)	82	82	100.0

→ Classification accuracy was concentrated on the Conventional (C) type group, while classification performance for the other types was low.

Table 3 Discriminant function coefficients and variable contributions by career type

Career type code	Energy direction	Information gathering	Decision making	Lifestyle
0 (R)	0.07	0.01	−0.43	−0.28
1 (I)	−0.23	−0.13	0.67	−0.19
2 (A)	1.11	0.21	0.75	0.24
3 (S)	−1.08	−1.52	0.86	0.61
4 (E)	0.09	−0.94	1.67	1.05
5 (C)	0.04	0.21	−0.28	0.01

→ Decision Making (F/T) and Lifestyle (Z/A) dimensions howed particularly high influence in Artistic (A), Social (S), and Enterprising (E) types, suggesting that individual tendencies regarding value judgment and organizational style play crucial roles in career selection

3.4.4 Comprehensive Summary of Research Results

This study aimed to classify the career types of cyber university students based on the four dimensions of the GOLDEN Personality Type Profiler and derived the following conclusions. First, an overall classification accuracy of 53.95% suggests that personality dimensions alone can predict career types to some extent, highlighting the practical applicability of a personality-based approach in career counseling scenarios. Second, the discriminant model demonstrated particularly high accuracy in predicting the Conventional (C) career type, indicating a potential relationship with personality characteristics emphasizing structural and organizational traits. Third, among the four personality dimensions, Decision Making and Lifestyle dimensions showed relatively higher influence in differentiating career types, suggesting that personal value judgments and working style preferences may serve as significant criteria in career selection. Lastly, although statistical significance was limited ($p = 0.081$), this exploratory analysis underscores the potential of personality dimension-based career prediction models.

4 Discussion

The results of this study significantly demonstrate the structural relationship between the personality characteristics and career types of cyber university students. Notably, among the four psychological dimensions of the GOLDEN personality test, the Decision Making (Thinking/Feeling) and Lifestyle (Judging/Perceiving) dimensions showed the most substantial influence on career types. This indicates that personality types transcend mere personal inclinations and directly connect to essential career preference elements such as job-environment congruence, criteria for value judgments, and approaches to work tasks. Specifically, Feeling (F) tendencies relate closely to people-oriented and value-oriented jobs such as Social (S) and Artistic (A), while Judging (Z) tendencies prefer structured environments and clearly defined rules found in Conventional (C) or Enterprising (E) career types. The integrated interpretation of GOLDEN and Holland career assessments provides a refined understanding of cyber university students' psychological tendencies and career preferences.

Cyber university students differ markedly from traditional university students in age, career experiences, and learning motivations. They engage in learning to achieve career transitions, self-development, and certifications, making career exploration a reconfiguration of current life rather than just future job selection. Therefore, the career and personality assessments applicable to these students should be realistic and appropriate for adult developmental stages.

The GOLDEN personality test comprehensively captures clients' behavioral tendencies, information processing methods, decision-making styles, and daily rhythms through its refined four dimensions (Energy Direction, Information Gathering, Decision Making, Lifestyle). The Decision Making and Lifestyle dimensions, particularly influential in this study, are directly relevant to actual job performance. Feeling types (F) prioritize value-oriented decisions, and Judging types (Z) favor clear planning and systematic task execution. These characteristics align closely with Holland's Conventional (C) and Social (S) career types.

From the perspective of career types, Holland's RIASEC model is designed around congruence between personality types and occupational environments, providing grounds to interpret compatibility between personality traits and preferred career environments among cyber university students. For example, most respondents classified as Conventional (C) types in this study exhibited preferences for organizational stability and procedure-oriented work environments, aligning with GOLDEN's Judging (Z) and Thinking (T) tendencies.

This study underscores the necessity of integrating personality assessment results when applying career assessments to cyber university students. Interest-based career assessments alone are insufficient for explaining these students' career preferences; information from personality dimensions allows for more precise predictions of career types when combined with practical elements such as decision-making methods, organizational adaptability, and interpersonal orientation.

These findings provide a theoretical foundation for actively employing GOLDEN assessment results in career counseling analyses. Particularly for adult learner groups such as cyber university students, the study offers significant insights for career counseling, career coaching, and Human Resource Development (HRD) program development. Practically, the results suggest possibilities for tailored career exploration strategies based on GOLDEN personality dimensions. For instance, cyber university students with Feeling (F) tendencies likely pursue relationship-oriented and value-driven careers, thus requiring counseling interventions that consider these characteristics. For these students, criteria like interpersonal quality, organizational culture harmony, and social significance might be crucial in job selection, making in-depth interventions such as value exploration and emotional strategies during career decisions highly effective.

Furthermore, there is potential for developing career coaching models centered on the GOLDEN personality assessment. The GOLDEN dimensions, moving beyond simple explanations of tendencies, offer a structure capable of predicting practical behavioral patterns and occupational environment preferences. This applicability extends to HRD programs, career design workshops, and career coaching model developments targeting adult learners, including cyber university students. For example, learners with Judging (Z) tendencies may benefit from structured career path settings and long-term goal planning, while those with Intuition (N) tendencies might find career designs emphasizing creativity and possibilities more suitable. This study utilized the GOLDEN personality test and Holland career type assessment among cyber university students, exploring through discriminant analysis the extent to which personality dimensions effectively classify career types. The discussion can be summarized as follows:

First, the overall classification accuracy of 53.95% indicates that stable psychological traits such as personality dimensions alone hold predictive potential for career types. This provides grounds for utilizing personality assessment results during initial diagnosis phases in career counseling, particularly valuable for cyber university students who differ from traditional career exploration subjects due to their self-directedness and career transition intentions.

Second, the concentrated analysis results on Conventional (C) types require further interpretation. Many cyber university students might prefer job stability and administrative-organizational environments, aligning with GOLDEN personality characteristics emphasizing structure and predictability. The relatively higher influences of Lifestyle (Z/A) and Decision Making (F/T) dimensions suggest that organizational and interpersonal values significantly impact career preferences. These results support existing studies indicating that individuals with Feeling and Perceiving tendencies may prefer stable, procedure-oriented job environments over Social (S) or Artistic (A) roles.

Third, the highly skewed classification accuracy across career types represents both analytical limitations and realistic reflections. The unbalanced distribution of career types in the sample and the inadequacy of personality dimension combinations to fully explain vocational aptitudes highlight the need to consider additional variables (e.g., academic departments, age, job experience). Future research should adopt multidimensional discriminant models, cluster analyses, or decision-tree methods to enhance explanatory power.

Lastly, this study's practical examination of the GOLDEN personality assessment's applicability holds significance. While most career studies utilize MBTI, GOLDEN offers a more structured dimensional interpretation and greater organizational applicability, making it particularly valuable for adult learners and career developers. This research significantly contributes to validating and enhancing the effectiveness of GOLDEN-based career diagnostics in adult career counseling, HRD, and coaching contexts.

5 Conclusion

This study has similarities and differences when compared to previous research. Existing studies [7] 1 primarily focused on the correlation between MBTI personality types and Holland's RIASEC career types, mostly using traditional university students as samples. In contrast, this study is distinctive both in terms of its sample characteristics—adult cyber university students—and its use of the more structured and sophisticated GOLDEN personality assessment. Furthermore, by employing discriminant analysis rather than simple correlation analysis, the study advanced the analytical framework, empirically confirming the predictive and classificatory potential of personality dimensions. Consequently, career type classification based on personality dimensions was possible to a certain degree, especially highlighting the significant influence of Decision Making and Lifestyle dimensions. The high predictive accuracy for the Conventional career type underscores the practical feasibility of career planning grounded in the personality-career congruence theory.

The study carries theoretical and practical significance. First, it contributed to expanding and diversifying the scope of career psychology research by building a career type prediction model around atypical learners—cyber university students. Second, it demonstrated the possibility of using the GOLDEN personality test, offering a more structured and practical approach compared to MBTI, thus presenting an alternative to traditional career assessment tools. Third, by providing objective evidence through the advanced statistical method of discriminant analysis beyond simple correlations, it enhanced the precision and validity of career counseling.

However, the study has several limitations. First, the concentration of career types on the Conventional (C) type reduces precision in comparisons across types, potentially restricting the generalizability of the results. Second, by using personality types alone as independent variables, it overlooked complex factors such as social-environmental influences, departmental characteristics, and career paths that might impact career types. Third, the discriminant analysis used had somewhat limited statistical significance, indicating the need for future adoption of diverse analytical techniques such as structural equation modeling or machine-learning-based classification models.

To overcome these limitations, future research could consider the following directions. First, obtaining a balanced distribution of career types and expanding the sample of cyber university students across various age groups and academic disciplines would enhance external validity. Second, incorporating additional psychological variables related to careers, such as career barriers, career beliefs, and career decision self-efficacy, could improve explanatory power in predicting career types. Third, empirical research should be conducted concurrently to establish interpretation guidelines and intervention protocols for utilizing the GOLDEN assessment in career counseling and coaching settings.

In summary, this study holds significant theoretical and practical implications by empirically demonstrating the structural relationship between personality types and career types among the specialized group of cyber university students, thereby presenting possibilities and directions for personality-based career counseling. It empirically analyzed the structural relationship between the personality types of cyber university students (GOLDEN assessment) and career types (Holland assessment) through discriminant analysis. Results confirmed the capability to classify career types based on personality dimensions to some extent, especially highlighting the significant influence of Decision Making and Lifestyle dimensions. The high predictive accuracy for the Conventional career type supports the practical feasibility of personality-career congruence-based career planning.

The implications of this study include the following: first, by constructing a career type prediction model around atypical learners such as cyber university students, it secured practical extensibility of its target population. Second, using the GOLDEN personality test instead of MBTI, it proposed a more structured and practical possibility for career analysis. Third, beyond simple correlation analysis, it provided objective evidence on the feasibility of predicting career types through discriminant analysis.

Future research should address the balance of career type distributions within samples and include various variables such as academic departments, career maturity, job experience, and career barriers to improve the precision of career type prediction models. Additionally, follow-up studies are required to systematize interpretation guidelines and career intervention strategies for practical application of the GOLDEN assessment in career counseling and coaching.

References

Kim MR (2020) Development of counseling programs for career transition among adult learners. Lifelong Educ Res 26(4):35–57

Golden L (2009) Golden personality type profiler: user's guide. Pearson

Savickas ML (2005) The theory and practice of career construction. In: Brown SD, Lent RW (eds) Career development and counseling: putting theory and research to work. Wiley, pp 42–70

Super DE (1957) The psychology of careers. Harper & Row

Dawis RV, Lofquist LH (1984) A psychological theory of work adjustment: an individual-differences model and its applications. University of Minnesota Press
Park SH, Kim JH (2017) Exploratory research on the relationship between personality types and career aptitude. Korean J Counsel 18(3):121–139
Holland JL (1997) Making vocational choices: a theory of vocational personalities and work environments, 3rd edn. Psychological Assessment Resources

The Competency Model Framework for Open Source Software Developers

Jong-Bae Kim, Sangah Kim, and Sungtaek Lee

1 Introduction

Open Source Software (OSS) refers to software in which the copyright holder discloses the source code, granting permissions for its free use, reproduction, modification, and distribution. Over the past few years, OSS has fundamentally transformed traditional software industry business models, which were previously reliant on intellectual property and software license sales. Today, OSS stands as a core component of strategic planning for major IT corporations.

Notably, OSS is not merely a specific type of software; it represents software in its entirety. The term "software developer" broadly includes open-source software developers, reflecting the integrated nature of OSS within software development as a whole. Consequently, OSS competencies encompass standard software development skills while requiring additional capabilities tailored to open-source environments.

This study does not distinguish OSS developers as a separate category but instead defines all developers employing open-source methodologies as OSS developers. Rooted in the principles of source code transparency and open participation, the open-source development model requires prospective developers to understand and acquire specific competencies (knowledge and skills). OSS competencies can be defined as measurable behavioral attributes exhibited when individuals effectively

J.-B. Kim
Startup Support Foundation, Soongsil University, Seoul, Republic of Korea
e-mail: kjb123@ssu.ac.kr

S. Kim
Department of IT Policy and Management, Graduate School of Soongsil University, Seoul, Republic of Korea
e-mail: ddingyo@na.go.kr

S. Lee (✉)
Department of Business Administration, Yong In University, Gyeonggi-Do, Republic of Korea
e-mail: totona22@yiu.ac.kr

R. Lee et al. (eds.), *IT Service & Business Innovation*, Studies in Computational Intelligence 1252, https://doi.org/10.1007/978-3-032-14397-6_4

combine knowledge, technical skills, and attitudes to excel in open-source roles or tasks.

This research proposes a systematic framework to define and model the competencies essential for software developers in OSS environments. Beyond technical expertise, OSS demands collaborative skills and community-driven communication abilities, making a structured competency model indispensable for evaluation and development.

The study identifies and categorizes the required competency elements through comprehensive literature analysis. This framework facilitates self-assessment of competency levels and enables individuals to establish structured development roadmaps. Furthermore, it is designed to support companies and educational institutions in creating tailored training programs and recruitment criteria aligned with the needs of the OSS ecosystem.

By addressing the critical competencies necessary for OSS environments, this study contributes to enhancing individual capabilities, promoting sustainable growth in OSS, and driving innovation across the broader software industry.

Some researchers argue that competencies are neither innate nor naturally developed over time. Instead, competencies are cultivated and enhanced throughout an individual's lifetime through continuous interaction with their environment.

According to Kaufhold, competencies are inherently tied to the individual, meaning they can be developed directly by the person or in response to various external demands from third parties [1].

Grundmann emphasizes that there are differences among competencies in terms of trainability; according to him, some competencies, which are primarily composed of knowledge, are highly trainable, whereas others may be less or only partially trainable due to their association with personal attributes [2].

Ultimately, competencies are not innate but can evolve through individuals' interactions with social realities. Trainable competencies can be developed independently by individuals or continuously improved throughout life in response to various external demands.

Therefore, this study aims to provide clear standards for the competencies required of developers within the OSS ecosystem. By doing so, it can guide developers in enhancing their OSS-related competencies, thereby contributing to their sustainable personal growth and, more broadly, to innovation across the industry.

2 Related Works

Research on competencies, which originated in the field of psychology, has continued over the past several decades and has since expanded into various practical domains. However, a standardized definition of competency has yet to be established [3].

The term competency originates from the Latin word "competentia", which means "having the authority to judge" or "having the right to speak." Historically, it was also used to describe the behavior of those in teaching or authoritative roles. A related concept is ability or capability, which may be either learned or innate and focuses on the question of "what one can do [4]." However, this concept differs somewhat from the notion of competency as defined in this study.

The concept of competency began to be used when early industrial psychologists sought to describe the characteristics of individuals who performed successfully in specific work settings. In the 1920s, Frederick Taylor argued that identifying and classifying competencies—that is, the components of specific tasks—should be a key responsibility of management scientists. He was among the first to use the term competency, proposing that work be broken down into detailed elements [5].

In 1954, psychologist John Flanagan introduced the concept of competency through the development of the Critical Incident Interview technique as part of the U.S. Air Force's Aviation Psychology Program. He described this technique as a systematic method for directly observing and collecting human behaviors, with the aim of solving real-world problems and developing broad psychological principles [6].

Subsequently, psychologist White was one of the first to conceptualize the notion of competency. He introduced the term in the context of personal characteristics, defining competency as the ability to interact effectively with one's environment. According to White, competencies are not driven solely by instincts or desires but are learned attributes that involve selective, sustained, and motivated behaviors aimed at managing one's surroundings. In essence, he emphasized that competencies are not innate but can be developed through learning [7].

McClelland introduced the concept of competency as a new conceptual approach distinct from the widely used notion of intelligence at the time [8, 5]. While acknowledging that intelligence can influence performance, he argued that personal characteristics—such as one's motivations or self-image—are the key factors that differentiate successful performance and significantly affect individual roles [8]. He defined competency as an intrinsic characteristic of an individual that leads to effective or superior performance and conducted research to identify such competencies among U.S. State Department Foreign Service information officers. In his seminal paper, "Testing for Competence Rather Than Intelligence," McClelland proposed a competency-based approach to personnel selection, arguing that competencies, rather than intelligence, should serve as the primary criteria since they more directly impact job performance [5]. This behavioral approach became foundational in the development of competency models and was later established as a methodology that incorporates not only observed behaviors but also individuals' experiences and perceptions of events.

The researcher who expanded competency studies into the managerial domain was Boyatzis. In his book, The Competent Manager, he discussed the essential (latent) characteristics causally related to superior performance, marking the beginning of practical interest in competencies [9]. In his study on managerial competencies, he observed over 2,000 managers across 12 organizations to identify differences between high performers and average performers. He defined competency as an internal characteristic that leads to effective or outstanding performance and collected and analyzed data to uncover the traits possessed by high performers. Boyatzis categorized competencies into motivation and traits, self-image and social roles, and skills, attempting to understand individual behavior comprehensively. As a result, he identified six competency clusters and 21 specific competencies. By emphasizing the intrinsic aspects of individuals that influence job performance, he highlighted psychological dimensions that had been overlooked in previous competency models.

Subsequently, Spencer [10] built upon the research of McClelland [8] and Boyatzis [9]. They defined competency as an individual's internal characteristics that serve as the underlying cause enabling effective and superior performance in specific situations or jobs [10]. Internal characteristics refer to relatively enduring patterns of behavior and thinking that generally manifest across various situations. They categorized the elements that constitute competencies into five groups based on their visibility: motives, traits, self-concept, knowledge, and skills. Motives are the internal drives that cause behavior. Traits are physical characteristics that result in consistent responses to certain situations or information. Self-concept encompasses attitudes, values, or self-image. Knowledge refers to detailed information about a specific area, such as a surgeon's understanding of the nerves and muscles in the human body. Skills denote the ability to perform given mental or physical tasks.

Synthesizing the definitions of competency proposed by various scholars, it can be observed that until the 1990s, the approach to competency primarily focused on the individual level. From the late 1990s onward, research began to explore competencies applicable at the organizational level. Entering the 2000s, the interest in competency expanded further, moving beyond individual or organizational performance to embrace a more systemic perspective that comprehensively addresses diverse environments and tasks.

In conclusion, competency refers to the characteristics manifested through behaviors necessary to achieve high performance at both the individual and organizational levels. These characteristics can be expressed through values, knowledge, skills, and attitudes, and collectively, through their interactions, constitute the total set of traits that drive successful outcomes for individuals or organizations.

Meanwhile, the concept of competency has been increasingly emphasized in the workplace, leading to the development of various competency models aimed at identifying essential competencies required of employees in organizations.

However, research that defines competencies and proposes models specifically tailored to the software (SW) field has largely been limited to elementary, secondary, and university education levels, showing limited scalability to the corporate environment. Moreover, competency model research from a corporate perspective has primarily focused on leadership development, with limited comprehensive studies addressing the entire career lifecycle of software developers through an integrated SW competency model.

Conger and Fulmer emphasize the importance of a pipeline that links continuous competency development with succession planning in a staged manner to cultivate leaders within organizations [11]. In their study, Lim Taejo and Lee Hyuneung introduced GE's leadership pipeline model and applied it to organizational leadership at various hierarchical levels; however, their approach did not extend to specific job roles and was limited to employees within the corporation [12]. Kim Suhwan et al. proposed a next-generation software (SW) education standard model consisting of eight stages covering elementary, middle, and high school curricula [13]. Additionally, Lee Hyejung et al. suggested comprehensive reforms and innovations in higher education based on the case of Minerva Schools in the U.S., aimed at cultivating talent demanded by society in the era of the Fourth Industrial Revolution [14]. Cha Joonseop conducted a survey of software courses in four-year universities and investigated corporate perspectives on university-level SW education [15].

Meanwhile, as artificial intelligence (AI) technology drives industrial transformation, there is increasing emphasis on the need to design AI literacy education models that enable university students to experience AI alongside cultivating their humanities-based competencies [16, 17]. In addition, the necessity of developing AI curricula grounded in computational thinking has been highlighted. Rivera-Ibarra et al. categorized the primary roles of software developers into ten types and proposed a four-level proficiency framework for the knowledge areas required in the field of software engineering [18].

In the United States, research on the comprehensive knowledge framework of the software (SW) field has been relatively continuous and active. Since 1968, the Association for Computing Machinery (ACM) and the Institute of Electrical and Electronics Engineers (IEEE) have collaboratively published university software education curriculum guidelines approximately every ten years. The CS Curricula 2013 involved over 130 experts and faculty members, defining 18 knowledge areas that should be covered at the university level, and categorizing them into three tiers based on importance: Tier-1, Tier-2, and Elective [19]. However, since these guidelines primarily target university-level SW courses, there are limitations in their immediate applicability for corporate use.

Additionally, the IEEE Computer Society has published the Software Engineering Body of Knowledge (SWEBOK), which consolidates the knowledge framework for the software engineering field by defining 15 knowledge areas. However, SWEBOK focuses specifically on software engineering and does not encompass the entire software domain [20].

Park Beomjoo et al. proposed a technical leadership competency model centered on cases from IT companies and highlighted the need for customized models tailored to specific technical fields, such as the software domain [21].

3 Open Source Competency Model Framework

3.1 Overview of Open Source Competency

In the rapidly growing field of open source software development, a diverse range of competencies is required, which are crucial for successful career management as an open source software developer [22].

Therefore, open source competency refers to measurable behavioral characteristics (elements) that emerge when knowledge, skills, attitudes, and various other factors are effectively integrated to achieve high performance in open source-related tasks or roles. These competencies have a decisive impact on accomplishing objectives through open source-related jobs or community activities.

First, open source competency is a behavior-centered concept; it is not simply the knowledge or skills an individual possesses, but the actions that result from the combination of inner motivation, values, attitudes, and knowledge. Therefore, it refers to the practical ability manifested through behavior rather than mere possession of capabilities.

Second, competencies are behaviors linked to performance; even if the behavior does not directly produce results for a specific job, it can be considered performance if it contributes to the advancement of the project. Competencies can be objectively assessed based on their contribution to performance.

Third, competencies consist of observable behaviors that can be measured objectively, and the measurement results should be able to provide feedback based on different proficiency levels.

3.2 Scope and Elements of Open Source Competency

The scope of open source competency can be divided into technical competencies, which involve utilizing and developing open source to produce new software, and activity competencies, which pertain to participating in open source communities as contributors or committers.

Among these, technical competencies include:

1. Understanding of open source concepts, licenses, and governance;
2. Familiarity with open source-based technologies and development environments such as Linux OS, IDEs, virtual machines, and containers;
3. Knowledge of programming languages, debugging tools, build tools, version control, and remote repositories;
4. Understanding of project execution and development methods, including source code integration and configuration management, issue tracking and management, source code repository and hosting environment setup, and packaging;
5. Comprehension of open source business models.

These technical competencies can develop through stages ranging from basic technical skills and application abilities to simple software production, software redesign, and advanced software production proficiency.

Next, activity competencies refer to actions performed within open source communities based on an individual's skills and authority, such as defect detection and removal, and feature enhancement. These activities include:

1. Patch activities such as debugging, refactoring, and developing new features;

2. Code review activities including bug tracking and suggesting improvements;
3. Documentation tasks such as typo correction, writing documents, creating tutorials, and translations.

When systematized further, these competency elements can be broadly classified and defined into the following five key categories.

(1) (Knowledge & Skills) Utilization of Collaboration Tools

The open source development approach involves publicly releasing software source code and allowing anyone, not just the original developers, to freely participate in the entire process of software development and maintenance. Regarding collaboration tools, source code is typically made accessible through public repositories that are easy to access and facilitate basic source code management. Among these, GitHub is the most widely used platform globally for managing open source projects. Understanding and proficiency in using such collaboration tools are essential competencies.

(2) (Knowledge) License Compliance

Open source software is protected under intellectual property rights, even though its source code is publicly available for anyone to freely use, copy, modify, and redistribute. Therefore, it is essential to understand license compliance, which specifies the scope of usage methods and conditions agreed upon between software developers and users.

(3) (Knowledge & Activity) Community

By its nature, open source involves publicly sharing source code and relies heavily on active participation from developers and users through vibrant communities to improve and advance the software. These communities serve as spaces to learn new technologies and meet new people. However, for newcomers to programming, open source communities can feel unfamiliar and intimidating. Therefore, understanding the open source community is essential, and ongoing participation in community activities is required for continuous growth as a developer.

(4) (Knowledge & Skills) Documentation

One of the biggest challenges in using open source is insufficient documentation. Well-prepared documentation is crucial for creating popular open source projects and communities, as it enables a broader range of people to participate effectively.

(5) (Knowledge & Activity) Business

Unlike free software (the early form of open source) which was perceived as entirely free, open source today encompasses various business models aimed at commercialization. As revenue-generating opportunities using open source have emerged, the spread of open source has accelerated, and the number of companies utilizing it has increased. Therefore, understanding open source business models is essential.

Based on the competency scope and elements described above, the open source competency components are redefined as shown in Table 1.

Table 1 Open source competency components

Category	Description
Development approach (Culture)	Strengthening open source software (OSS) competencies begins with understanding the development approach and culture that share the concepts, values, and philosophy of OSS. It involves recognizing the development process where the source code is publicly available, and anyone—not just the original developer—can freely participate in the development and maintenance of the software throughout its entire lifecycle
Utilization of collaboration tools	In typical open source software (OSS), the source code is made publicly accessible by selecting one of the public repositories that allow easy access and basic source code management GitHub is the most representative collaboration tool globally, managing the largest number of OSS projects. Therefore, understanding and proficiency in using such collaboration tools and methods are essential
License compliance	Open Source Software (OSS) discloses its source code, allowing anyone to freely use, copy, modify, and redistribute it; however, it remains a copyrighted work protected by intellectual property rights Therefore, it is necessary to understand license compliance, which specifies the scope of usage methods and conditions between software developers and users
Community	Due to its nature, Open Source Software (OSS) discloses its source code and relies on the active participation of many individuals. It is improved and advanced through the engagement of a vibrant community of developers and users The community serves as a space to learn new technologies and connect with others; however, for those new to programming, the OSS community can feel unfamiliar Therefore, understanding the OSS community is essential, and active participation is required for continuous growth as a developer
Documentation	One of the biggest challenges in utilizing Open Source Software (OSS) is the lack of sufficient documentation. Well-written documentation is essential for building successful OSS projects or communities, as it enables broader participation from diverse contributors. Creating high-quality documentation is therefore crucial
Business	Unlike Free Software—an early form of Open Source Software (OSS) that was commonly perceived as free of charge—modern OSS has evolved to include a variety of business models aimed at commercialization As revenue-generating strategies based on OSS have emerged, the adoption and proliferation of OSS have accelerated, with more companies leveraging OSS in their operations. Therefore, understanding OSS business models is essential

Table 2 Levels of open source community participation competency

Stage	Category	Definition
Stage 4	Maintainer	An individual who sets the direction of the project and manages it. This role is typically held by some of the committers
Stage 3	Committer	A person who reviews contribution content and decides whether to incorporate it into the project Commits changes to a central repository such as Git
Stage 2	Contributor	Anyone who makes contributions; contributions refer to all activities involved in participating and contributing to an open source project Types of contributions include: bug reporting and fixing (30%), documentation (28%), adding new features (18%), refactoring (9%), updating versions (7%), improving error handling (3%), optimizing resource usage (2%), and adding test cases (1%)
Stage 1	User	An individual who utilizes the outcomes or products of an open source project

3.3 *Open Source Competency Levels*

Meanwhile, the levels of open source competency can be graded (staged or leveled) and assessed (measured) from two perspectives: (1) skills (abilities) and knowledge, and (2) activities (participation in open source communities).

First, activity competency (the level of participation in open source communities) measures the ability to contribute to open source projects, specifically the ability to identify and fix source code defects and to enhance functionalities. This can be defined and assessed in four levels as shown in Table 2.

Next, the technical & knowledge competencies can be measured according to levels of open source technical skills, as shown in Table 3, which defines five levels.

3.4 *Open Source Competency Model Framework*

In summary, the Open Source Software (OSS) Competency Model Framework consists of five progressive levels of OSS competency: "Infrastructure/Tool Usage → Source Code Utilization → Community Participation → Project Contribution → Project Initiation/Leadership.“ These levels are mapped to six OSS competency components—Development Methodology (Culture), Collaboration Tools, License Compliance, Community, Documentation, and Business—and three competency characteristics—Knowledge, Skills, and Activities. Accordingly, the achievement goals (levels) for each stage can be defined and measured (diagnosed) as shown in Table 4.

Meanwhile, to quantify competencies, six measurement methods are proposed, and the measurement criteria for each method are presented in Table 5.

Table 3 Levels of open source technical (& knowledge) competence

Stage	Level	Definition
Stage 5	Skilled SW production	Develop new programs tailored to user requirements Transform key processes: Redesign of the software production process Process innovation: Designing entirely new software development processes
Stage 4	SW redesign	The capability to modify not only the software product developed through open source but also the underlying production processes
Stage 3	Simple SW production	Develop programs that enable data exchange between application software Modify software to meet user requirements Integrate open source software with other open source projects or proprietary software
Stage 2	Adaptation without production	Develop application software based on packages Use external code by importing it
Stage 1	Non-production technical capabilities	Awareness of open source, interest, and utilization of ideas Use of menu systems Ability to teach software usage, install software, and troubleshoot issues

These six methods are all direct or indirect ways to measure a developer's skills, and their relative importance may vary depending on the situation.

For example, participation in conferences, study groups, or obtaining relevant technical certifications are not directly reflected but can be recorded as activities performed during the process of acquisition.

Also, evaluation methods like interviews are generally excluded due to constraints such as time, location, and personnel, but organizations may choose to include them.

Finally, these six evaluation methods can be weighed appropriately according to the characteristics of roles, development teams, and projects.

Table 4 Open Source Software (OSS) Competency Model Framework

Competency stage(Level)	Infrastructure/tool usage	Source code utilization	Community participation	Project contribution	Project initiation/ leadership	Competency characteristics		
Competency element	Level 1	Level 2	Level 3	Level 4	Level 5	Knowledge	Skill	Activity
Development approach (Culture)	★☆☆☆☆	★★☆☆☆	★★★☆☆	★★★★☆	★★★★★	✓		
Utilization of collaboration tools	★☆☆☆☆	★★★☆☆	★★★★☆	★★★★☆	★★★★★	✓	✓	
License compliance	★☆☆☆☆	★★★☆☆	★★★☆☆	★★★☆☆	★★★★★	✓		
Community	★☆☆☆☆	★★☆☆☆	★★★☆☆	★★★★☆	★★★★★	✓		✓
Documentation	★☆☆☆☆	★★☆☆☆	★★☆☆☆	★★★☆☆	★★★★★	✓	✓	
Business	★☆☆☆☆	★☆☆☆☆	★☆☆☆☆	★★★☆☆	★★★★★	✓		✓

Table 5 Measurement methods for quantifying competencies

Category	Measurement method
Examination	• A single test is conducted for each technical unit • Test questions are selected from a question bank with a specified ratio of beginner, intermediate, and advanced level items • Each grade has a target score, which can be adjusted to reflect the characteristics of the organization • For example, the general target score for an intermediate server developer in the Java programming technical unit is 80 points, but an organization can adjust this target to 70 points
Project	• The project participation period using the specific technical unit is represented in months along with the project name • When multiple technologies are used together in one project, the development period for each technology is calculated separately • Then, career density is calculated and reflected in the overall career length. For example, if someone participates in a 12-month project using Java, iBATIS, and Struts, counting each as 12 months sums up to 36 months of experience. However, the career density is only 30%. In this case, the overall career experience is adjusted downward proportionally based on this career density
Reading	• The number of books read related to each technical unit is recorded • When a single book covers multiple technologies, the count is distributed proportionally, for example, 0.3 books, 0.5 books, etc • The title of each book must be listed • Like project experience, when one book includes several technologies, the reading density may be low. In such cases, the total number of books read is adjusted proportionally to reflect this
Lesson	• When participating as a student in lectures focused on or related to a specific technical unit, the participation is measured by the number of lecture hours attended • Like reading, if a course covers multiple technologies, the hours are recorded proportionally, and later the attendance density is reviewed to adjust the total lecture hours accordingly
Lecture	• When participating as an instructor in lectures focused on a specific technical unit, the teaching time is measured by the hours lectured • Like reading, if a course covers multiple technologies, the teaching hours are recorded proportionally, and later the teaching density is reviewed to adjust the total lecture hours accordingly
Publishing	• The number of published books or white papers focused on a specific technical unit is recorded as the count of those publications

4 Conclusion

This study classified the competency components required in the OSS environment and proposed a competency diagnosis framework necessary for individuals to assess their own competency levels and establish a systematic development roadmap.

By providing clear criteria for developer competencies required in the OSS ecosystem, this study aims to contribute to the sustainable development of OSS and to promote innovation across the industry.

However, the competency model presented in this study is at the framework level and applying it to measure OSS competencies of developers in the field requires the development of detailed diagnostic (measurement) indicators.

These detailed diagnostic indicators may vary depending on the characteristics of the organization and the purpose of competency measurement.

Nonetheless, the competency model framework presented in this study is expected to be useful for companies and educational institutions in developing customized training programs and hiring criteria tailored to the specific circumstances of each organization.

References

1. Kaufhold M (2006) Kompetenz und Kompetenzerfassung: analyse und Beurteilung von Verfahren der Kompetenzerfassung (1. Aufl). VS, Verl. für Sozialwiss, Wiesbaden. https://doi.org/10.1007/978-3-531-90444-3
2. Grundmann ST (2011) Making the right connections: targeting the best competencies for training. A report to the president and the congress of the United States. https://www.mspb.gov/studies/studies/Making_the_Right_Connections_Targeting_the_Best_Competencies_for_Training_581608.pdf
3. Schippmann JS, Ronald AA, Battista M, Carr L, Eyde LD, Hesketh B, Kehoe J, Pearlman K, Prien EP, Sanchez JI (2000) The practice of competency modeling. Person Psychol 53(3):703–740. https://doi.org/10.1111/j.1744-6570.2000.tb00220.x
4. Caupin G, Knoepfel H, Koch G, Pannenbacker K, Perez-Polo F, Seabury C (2006) ICB-IPMA competence base line version 3. IPMA, Nijkerk. http://www.pm.org.ro/ICB-V-3-0/ICB-V-3-0.pdf
5. Lucia AD, Lepsinger R (1999) Art and science of competency models. Jossey-Bass/Pfeiffer, San Francisco, CA. ISBN: 978-0-787-94602-9
6. Flanagan JC (1954) The critical incident technique. Psychol Bull 51(1):327–358. https://doi.org/10.1002/9781118785317.weom090062
7. White RW (1959) Motivation reconsidered: the concept of competence. Psychol Rev 66(5):297–333. https://doi.org/10.1037/h0040934
8. McClelland DC (1973) Testing for competence rather than for intelligence. Am Psychol 28(1):1–14. https://doi.org/10.1037/h0034092
9. Boyatzis RE (1982) The competent manager: a model for effective performance. John Wiley & Sons, New York, NY. ISBN: 0-471-09031-X
10. Spencer LM, Spencer SM (1993) Competency at work: models for superior performance. John Wiley and Sons Inc., New York, NY. ISBN: 978-0-471-54809-6
11. Conger JA, Fulmer RM (2004) Developing your leadership pipeline. Harvard Bus Rev 81(12):76–84, 125
12. Lim T, Lee H (2007) Leadership pipeline: application of the model and its relationship with performance. Korean Bus Educ Rev 46(1):243–259. uci: G704–001274.2007..46.015
13. Kim K, Koo D, Kim S, Kim S, Kim Y, Kim J, Kim J, Kim C, Kim C, Kim H, Kim H (2020) Development a standard curriculum model of next-generation software education. J Korean Assoc Inform Educ 24(4):337–367. https://doi.org/10.14352/jkaie.2020.24.4.337
14. Lee H, Im S, Kang S (2019) Implications for innovation in higher education from minerva schools. J Lifelong Learn Soc 15(2):59–84. https://doi.org/10.26857/JLLS.2019.5.15.2.59
15. Cha JS (2015) The study of the field customized SW training course design based on the analysis of the field suitability of the university SW education. Smart Media J 4(4):86–92. uci: G704-SER000003489.2015.4.4.006

16. Park YS, Yi Y (2021) The education model of liberal arts to improve the artificial intelligence literacy competency of undergraduate students. J Korean Assoc Inform Educ 25(2):423–436. https://doi.org/10.14352/jkaie.2021.25.2.423
17. Shin S (2019) Designing the instructional framework and cognitive learning environment for artificial intelligence education through computational thinking. J Korean Assoc Inform Educ 23(6):639–653. https://doi.org/10.14352/jkaie.2019.23.6.639
18. Rivera-Ibarra JG, Rodriguez-Jacobo J, Fernandez-Zepeda JA (2010) Competency framework for software engineers. In: 2010 3rd IEEE conference on software engineering education and training, pp 33–40. https://doi.org/10.1109/CSEET.2010.21
19. Computer Science Curricula (2013) The joint task force on computing curricula association for computing machinery(ACM). IEEE Computer Society 2013. https://doi.org/10.1145/2534860
20. IEEE Computer Society (2014) Guide to the software engineering body of knowledge version 3.0. Retrieved from https://www.computer.org/education/bodies-of-knowledge/software-engineering
21. Park BJ, Lee HJ, Shin WS (2015) A study on developing competency model of techno leadership. Korean Manag Sci Rev 32(2):1–14. https://doi.org/10.7737/KMSR.2015.32.2.001
22. Kimmelmann N (2013) Career in open source? Relevant competencies for successful open source developers. IT Inform Technol 55(5). https://doi.org/10.1524/itit.2013.1009

Big Data's Role in Structuring Youth Consumption Culture Within AI-Mediated Virtual Spaces

Jinhee Lee and Jinhee Son

1 Introduction

The rapid evolution of digital technologies has profoundly reshaped the everyday lives of youth, altering the ways in which they communicate, socialize, and consume. Among the most significant developments is the rise of AI-driven virtual spaces—interactive environments such as metaverse platforms, algorithm-based social media, and gamified ecosystems—that provide immersive, personalized, and increasingly commercialized experiences (Van Dijck and Poell 2013). These environments, underpinned by big data analytics, function not merely as passive venues of interaction but as dynamic systems that actively shape user behavior and cultural norms (Striphas 2015; Abidin 2020). Big data plays a central role in these transformations (Couldry and Mejias 2019). By collecting and analyzing extensive volumes of user-generated data—ranging from click patterns and location data to interaction histories and biometric signals—platforms construct detailed digital profiles. These profiles enable the delivery of hyper-personalized content, targeted advertisements, and algorithmically curated recommendations that significantly influence users' consumption behaviors and social preferences (Eubanks 2018). For youth, who are in the critical stages of forming identity, values, and social affiliations, such data-driven interventions have considerable implications for their sense of self, belonging, and aspiration (Papacharissi 2015).

J. Lee
Department of Finance Investment Technology, Korea Soongsil Cyber University, Seoul, Republic of Korea
e-mail: clara@mail.kcu.ac

J. Son (✉)
Department of Youth Coaching and Counseling, Korea Soongsil Cyber University, Seoul, Republic of Korea
e-mail: ssony@mail.kcu.ac

R. Lee et al. (eds.), *IT Service & Business Innovation*, Studies in Computational Intelligence 1252, https://doi.org/10.1007/978-3-032-14397-6_5

Youth consumption culture within virtual spaces is increasingly governed by opaque algorithmic logics (John 1999). Recommendation systems not only direct what users watch or purchase but also shape perceptions of desirability, trendiness, and social acceptability (Livingstone 2008). The algorithmic acceleration of viral trends contributes to a culture characterized by immediacy, novelty, and often superficial engagement with consumer goods and social norms (Marwick 2013). Of particular concern is the commodification of identity, whereby appearance, taste, and lifestyle are curated and optimized for digital visibility—an issue with growing relevance among adolescent users (Oh and Lee 2021). These developments also raise pressing ethical and sociopolitical questions. What values are embedded within the algorithms that filter and prioritize content? How do platform architectures nudge young users toward particular consumption behaviors, potentially reinforcing cultural stereotypes or excluding marginalized identities? What responsibilities do technology companies bear in shaping youth culture through data-driven design? While prior studies have examined youth participation in digital platforms, few have explored how algorithmic infrastructures not only reflect but actively construct youth consumer desire and identity within commercially driven digital ecologies. This study addresses this gap by critically examining the socio-cultural mechanisms through which big data systems influence youth consumption practices and identity formation. To this end, the study is guided by the following research questions:

RQ1. How do algorithmic recommendation systems influence the construction of youth consumer desire on TikTok, ZEPETO, and Roblox?

RQ2. In what ways does datafication reshape youth identity and value formation through personalized digital environments?

Through an interdisciplinary synthesis of literature from media studies, youth studies, and digital sociology, combined with case analyses of platforms such as TikTok, Roblox, and ZEPETO, this study investigates how data infrastructures, algorithmic mechanisms, and platform architectures govern youth behavior and socialization processes. Furthermore, it proposes a critical framework for understanding the socio-cultural implications of algorithmic personalization in shaping the values, aspirations, and identities of the younger generation. In doing so, the research contributes to ongoing scholarly debates on the ethical design of digital platforms, the cultural ramifications of big data, and the necessity of youth-centered policy interventions in the digital era. Ultimately, it calls for a more nuanced understanding of how data infrastructures intersect with youth agency, autonomy, and cultural participation in technologically mediated environments.

2 Theoretical Background

This study employs four interrelated theoretical frameworks to critically examine how big data technologies shape youth consumption culture in AI-driven virtual environments. The selected frameworks—algorithmic culture, digital consumer socialization, platform political economy and datafication, and the social construction of youth—collectively offer a comprehensive, interdisciplinary perspective through which to interpret the entanglement of algorithmic infrastructures with youth identity, behavior, and aspiration.

2.1 Algorithmic Culture

The concept of algorithmic culture, as proposed by Striphas(2015), reframes algorithms as cultural actors that co-construct meaning, social norms, and user behavior. Rather than functioning as neutral technological tools, algorithms organize attention, establish hierarchies of visibility, and encode normative standards. Gillespie(2014) describes these systems as "cultural editors," emphasizing their role in shaping what users perceive as desirable or relevant. Empirical studies demonstrate that platforms such as TikTok, Instagram, and ZEPETO operationalize algorithmic culture through engagement-based recommendation systems. These systems prioritize content that maximizes emotional resonance and platform retention. For instance, TikTok's "For You" page amplifies videos based on user interaction metrics such as watch time, comments, and replays. Research on Generation Z users suggests that such algorithmic filtering not only reflects interest but actively constructs patterns of taste and consumption, steering youth toward stylized influencers and branded aesthetics. This process is particularly formative during adolescence—a period marked by identity exploration and peer comparison (Erikson 1968). Algorithmic curation amplifies dominant discourses and commercialized ideals, functioning as a mechanism of cultural reproduction that privileges visibility-compatible identities over alternative expressions. In doing so, platform algorithms shape how youth learn to see themselves, what they emulate, and which performances are socially rewarded.

2.2 Digital Consumer Socialization

Consumer socialization theory, rooted in developmental psychology and consumer behavior (John 1999), originally emphasized the role of family, school, and peers in transmitting consumption-related knowledge and norms. In algorithmically mediated digital ecosystems, however, these roles are increasingly assumed by platforms themselves, which operate as pedagogical spaces of commercial learning. Contemporary platforms like Roblox, ZEPETO, and TikTok act as "algorithmic educators," embedding market-oriented values through personalized recommendations, social feedback mechanisms, and gamified incentives. These systems do not simply promote consumption; they socialize youth into consumer logics, teaching not only what to buy, but also why and how to consume within platform-specific cultural grammars. A key example is avatar customization, prevalent in Roblox and ZEPETO. While often framed as a form of creative expression, such customization is embedded in visibility hierarchies and monetization systems. According to the 2022 *Roblox Developer Report*, over 70% of teenage users engage in regular avatar modification, often influenced by algorithmically promoted digital goods. These practices are not merely aesthetic but constitute acts of stratified consumption, through which youth negotiate social status and digital belonging.

This aligns with the frameworks of Dittmar(2008) and Chan(2013), who conceptualize consumer-oriented identity construction as a process wherein self-concept becomes tied to visible markers of taste and brand affiliation. However, the digital socialization process is far from neutral; it tends to reinforce competitive, exclusive, and aspirational norms, marginalizing non-conforming youth and embedding consumption as a developmental imperative.

2.3 Platform Political Economy and Datafication

The notion of datafication, as introduced by Van Dijck and Poell(2013), highlights how everyday digital activity—likes, clicks, scrolls—is converted into quantifiable data for economic exploitation. In youth-oriented platforms, these activities are situated within a surveillance capitalist infrastructure (Zuboff 2019) that extracts behavioral data to refine predictive modeling and personalize engagement. This transformation produces a dual condition: youth are simultaneously producers of content and commodities of data. Every interaction contributes to a feedback system that shapes future content exposure, advertising segmentation, and even interface design. While this may appear seamless to the user, it fundamentally alters the logic of youth participation by embedding it in algorithmic value systems. A salient example is Instagram's algorithmic feed, which curates content based on prior behavioral patterns. Yet many adolescent users remain unaware of how their data is collected, processed, and repurposed—a disconnect that raises critical concerns about digital literacy, informed consent, and algorithmic transparency. Crucially, this political economy is not ideologically neutral. As Noble(2018) and Eubanks (2018) argue, algorithmic infrastructures frequently replicate and amplify pre-existing social inequalities. Content that aligns with dominant norms—around race, gender, class, and beauty—is more likely to be promoted, while alternative expressions are suppressed. Thus, platforms operate as soft governance systems, filtering which aspirations are rewarded and which identities are deemed visible or valuable.

2.4 Social Construction of Youth

The social constructionist view, articulated by James and Prout(1997), challenges essentialist narratives of adolescence as a biologically uniform stage. Instead, it conceptualizes youth as a discursively and structurally produced category, shaped by historical, institutional, and technological forces. In algorithmically governed digital contexts, youth identity is not merely self-generated but co-constructed through interactions with platform architectures. Platforms such as ZEPETO and TikTok function as socio-technical assemblages, where algorithmic visibility, monetization tiers, and design affordances collectively determine which identities are amplified and which are marginalized. For example, ZEPETO offers expansive options for avatar-based self-expression, but meaningful customization often requires financial investment—premium items, virtual currencies, or exclusive packages. This introduces a stratified system of self-representation, where the capacity to be seen or celebrated depends on economic means. These dynamics mirror and reproduce offline inequities in digital form.

From a constructionist standpoint, youth agency is always situated and negotiated. Navigating platform participation entails engaging with material limits (e.g., access to features) and normative scripts (e.g., platform-specific beauty ideals). This raises key theoretical questions:

Does digital self-expression on these platforms enable authentic identity formation, or merely incentivize conformity to algorithmically legible archetypes?

To what extent are youth genuinely empowered to construct identities, and where are they subtly nudged toward market-optimized personas?

Ultimately, this framework helps illuminate how platform-mediated youth cultures function both as sites of creativity and as mechanisms of regulation, entrenching particular forms of visibility, desirability, and participation within algorithmically governed environments.

3 Literature Review

Recent interdisciplinary scholarship has critically explored the evolving intersection of youth culture, algorithmic technologies, and digital consumption—particularly within immersive and AI-powered platforms. This section synthesizes key insights from media studies, youth studies, and digital sociology to conceptualize the mechanisms through which platform architecture, datafication, and algorithmic mediation shape young users' identities, behaviors, and values. Four strands of research are reviewed below.

3.1 Youth Engagement with Algorithmic Platforms

Numerous studies have examined how AI-driven platforms such as TikTok, Roblox, and ZEPETO are transforming youth participation in digital culture. These platforms function not just as tools for leisure but as algorithmically structured socio-technical systems that mediate identity formation and social interaction. As Montgomery(2015) notes, such platforms constitute a "new public sphere" for adolescents, where algorithmic metrics like likes, views, and shares increasingly determine visibility and cultural capital. Building on this, Livingstone and Blum-Ross(2020) argue that youth are not passive consumers but active meaning-makers, navigating within affordances shaped by platform design. However, these affordances are embedded in commercial logics that privilege conformity to dominant norms—particularly in appearance, gender, and lifestyle aesthetics.

Empirical studies, such as Nandagiri and Ghaffar(2022), demonstrate that youth often tailor their content (e.g., participating in viral trends, applying filters) not for self-expression but to meet algorithmic expectations. This illustrates the concept of algorithmic gatekeeping, wherein platform systems subtly guide expression through content visibility hierarchies. Notably, this process produces intersectional disparities: marginalized youth often face algorithmic underexposure, reinforcing systemic exclusions—a phenomenon Noble(2018) refers to as algorithmic exclusion.

3.2 Personalized Recommendation and Aspirational Consumption

Recent research highlights how personalized recommendation systems serve as cultural and emotional infrastructures, shaping how youth engage with content and internalize consumer norms. Unlike earlier recommendation tools, contemporary algorithms operate through behavioral profiling and real-time engagement tracking (de Reuver et al. 2020; Cotter 2019).

This is particularly consequential during adolescence, a stage marked by identity development and heightened peer sensitivity (Erikson 1968). Platforms such as TikTok and Instagram repeatedly expose users to curated portrayals of idealized lifestyles. These are not simply entertainment but function as aspirational scripts, promoting particular beauty standards, consumption behaviors, and influencer archetypes. Duffy(2017) introduces the concept of aspirational labor to describe the identity work involved in emulating successful creators. Bishop (2021) further conceptualizes this as reflexive consumption, where youth engage in consumption as a way of self-definition. However, this system also fosters algorithmic conformity—users adapt their preferences and behavior to fit platform logics that favor virality and trendability over authenticity or diversity. The unequal effects of this system are well-documented: youth from marginalized or lower-income backgrounds often feel pressure to adopt dominant aesthetics to gain visibility (Noble 2018; Bishop 2021). These findings, while empirically grounded, raise ethical concerns about emotional manipulation and algorithmic inequity in shaping youth consumption patterns.

3.3 Digital Identity Commodification and Peer Pressure

The commodification of youth identity on digital platforms is a growing concern in critical scholarship. Drawing from a political economy perspective, scholars argue that platforms convert identity into performative and monetizable capital (Marwick 2013). The notion of visibility labor captures the often-invisible emotional and cognitive work youth perform to remain algorithmically relevant. Marwick(2013) and Papacharissi(2015) describe platforms as affective infrastructures, designed to solicit engagement through likes, shares, and follower metrics. These metrics not only validate users socially but serve as behavioral regulators—shaping which performances are rewarded or marginalized.

Ethnographic and platform-based studies show how this results in a double bind: while platforms promise empowerment through self-branding, they simultaneously constrain authenticity by embedding youth identity within attention economies. Aesthetic coherence, trend alignment, and brand mimicry often take precedence over self-expression. Moreover, this dynamic is not experienced equally. Marginalized youth—such as those identifying as queer, non-white, or outside dominant beauty norms—are disproportionately subjected to algorithmic invisibility (Noble 2018; Bishop 2021). Thus, being seen becomes a contested and unequal process, mirroring broader social hierarchies within digital space.

3.4 Research Gaps: Datafication and Cultural Production

While existing literature offers rich insights into youth engagement, identity work, and platform governance, a significant theoretical and empirical gap remains: few studies fully address how big data analytics operate not just in the background, but as active cultural agents shaping youth consumption. Much research has either focused on macro-level critiques (e.g., surveillance capitalism, commodification) or micro-level psychological perspectives. As Zuboff (2019) and Couldry and Mejias (2019) argue, the commodification of user behavior through data infrastructures has broader cultural implications. However, youth studies have yet to fully theorize how datafication co-produces aspiration, desirability, and visibility within platform logics.

This study addresses this lacuna by integrating perspectives from digital sociology, youth development theory, and critical data studies, to reconceptualize big data not merely as a tool for personalization or prediction, but as a sociotechnical and ideological force. In doing so, it highlights how youth are not only shaped by algorithmic culture but also co-produce and internalize it through daily interaction with data-driven systems.

4 Methodology

4.1 Research Design

This study adopts a qualitative case study design to explore how big data and algorithmic infrastructures shape youth consumption culture within AI-driven virtual environments. A qualitative approach is particularly suited to uncovering the complex, situated, and socio-technical dynamics that characterize youth-platform interactions—dynamics that are often obscured by large-scale quantitative metrics yet deeply embedded in everyday digital practices. The case study methodology was selected for its capacity to provide context-rich and holistic insight into contemporary social phenomena, particularly those that are fluid, dynamic, and situated within specific cultural and technological environments (Yin 2018). In this research, the selected platforms—TikTok, ZEPETO, and Roblox—serve as instrumental cases, illustrating the multifaceted processes through which youth are socialized, surveilled, and shaped by algorithmic systems. Rather than seeking statistical generalization, the study prioritizes analytic generalization, wherein theoretical insights derived from in-depth case analysis are used to refine and extend existing conceptual frameworks (Flyvbjerg 2006). Drawing upon theories of algorithmic culture, consumer socialization, datafication, and the social construction of youth, this design facilitates interpretive depth. Moreover, this methodological framework supports the triangulation of diverse data sources—including platform policies, interface observations, user narratives, and media discourses—thereby enhancing rigor while capturing the nuanced and evolving experiences of youth within digital environments. The case study thus functions both as an empirical lens and as a theoretical bridge linking micro-level identity practices with macro-level platform governance structures.

4.2 Case Selection

A purposeful sampling strategy was employed to select three digital platforms—TikTok, ZEPETO, and Roblox—as representative cases for analyzing how big data and algorithmic systems mediate youth consumption culture in AI-based virtual environments. The selection was guided by both theoretical significance and empirical relevance (Patton 2015), with each platform exemplifying distinct yet complementary dynamics of youth digital engagement.

Three intersecting criteria informed the platform selection:

1. Global popularity among adolescents, ensuring relevance and cultural penetration;
2. Integration of algorithmic recommendation systems and datafication processes;
3. Facilitation of identity performance and performative consumption through platform-specific features.

TikTok represents the algorithmic virality model, with short-form videos curated via real-time machine learning that optimizes visibility based on behavioral data. It is a site of youth-driven trend performance, influencer culture, and aspirational consumption shaped by viral aesthetics and branded content (Abidin 2020). ZEPETO embodies avatar-based identity construction within a highly gamified and commercialized context. The platform's monetized personalization features promote alignment with digital fashion standards and signal social status within immersive virtual environments (Oh and Lee 2021). Roblox offers a user-generated platform economy, where youth create, customize, and monetize interactive experiences. Through its virtual goods system, Robux currency, and developer incentives, the platform supports entrepreneurial consumption and peer-driven cultural production (Nandagiri and Ghaffar 2022). Collectively, these platforms offer a strategically diverse analytical frame to interrogate the entanglement of youth, data infrastructures, and consumer socialization. Despite their differing formats, all three share reliance on personalized algorithms, gamified engagement, and behavioral feedback mechanisms, rendering them well-suited for comparative analysis within a qualitative case study framework.

4.3 Case Selection

To ensure methodological rigor and analytic depth, this study employed a multi-source qualitative data strategy rooted in principles of digital ethnography (Pink et al. 2016; Hine 2015). This approach recognizes the co-constitutive nature of user behavior, platform architecture, and cultural meaning in online spaces, necessitating layered and contextually situated inquiry. Over a six-month research period, data were collected from the following four categories:

(1) **Platform Ethnography**: Adapting ethnographic fieldwork to virtual environments (Boellstorff et al. 2012), the researcher conducted sustained engagement with TikTok, ZEPETO, and Roblox. This included daily exploratory use, observation of algorithmic responses, participation in content ecosystems, and systematic note-taking on interface dynamics, monetization structures, and emergent user practices.
(2) **Policy Documents and Platform Disclosures**: A critical documentary analysis was conducted on official platform texts, including privacy policies, terms of service, algorithmic transparency reports, and governance frameworks. These materials were analyzed to uncover institutional logics, economic imperatives, and the framing of responsibilities toward youth users.
(3) **User Narratives and Digital Artifacts**: User-generated content such as YouTube vlogs, TikTok comment threads, Reddit discussions, and Twitter/X posts—particularly those authored by self-identified adolescents—were examined to explore how youth interpret, experience, and negotiate algorithmic

systems. These narratives offered rich insight into youth agency, aspiration, resistance, and platform literacy.

(4) **Media Discourse and Expert Commentary**: To contextualize empirical findings within broader sociotechnical debates, the study reviewed news coverage, investigative tech journalism, and academic blog posts. These sources provided valuable commentary on controversies surrounding datafication, platform capitalism, digital rights, and algorithmic harms, helping situate youth experiences within a macro-level discursive frame.

Through the triangulation of these sources, the study captured both the lived experiences of users and the institutional-technical systems that structure them. This reflexive, multi-scalar approach aligns with emerging methodologies in digital ethnography that emphasize the infrastructural and experiential dimensions of algorithmic governance.

4.4 Data Analysis

Data were analyzed using thematic analysis (Braun and Clarke 2006), guided by an abductive reasoning approach that moved iteratively between empirical data and theoretical frameworks. Four primary thematic categories emerged, as outlined in the Table 1.

Coding was performed both manually and with the aid of qualitative analysis software (e.g., NVivo) to ensure consistency and transparency in thematic interpretation. Throughout the process, the researcher maintained critical reflexivity, attending to their own positionality, platform familiarity, and interpretive bias. This analytic strategy allowed for a nuanced exploration of how algorithmic infrastructures co-shape youth identities, consumption practices, and platform-based socialization.

Table 1 Thematic coding framework: youth engagement in AI-driven platforms

Theme	Examples
Identity	Avatar customization, personal branding, algorithmic exposure
Aspiration	Follower growth, monetization goals, role model emulation
Visibility	Algorithmic amplification, aesthetic trends, peer approval
Influence	Behavioral shifts, trend conformity, platform nudging

5 Findings

This section presents the core findings of the study, organized around three interrelated themes that emerged from the analysis of TikTok, ZEPETO, and Roblox. These themes illuminate how big data systems and algorithmic infrastructures shape youth consumption culture within AI-driven virtual environments.

5.1 Theme 1: Algorithmic Shaping of Desire

Across all three platforms analyzed—TikTok, ZEPETO, and Roblox—youth are embedded in algorithmically curated digital ecologies that do not merely respond to user preferences but actively construct desire. These environments function not as neutral mirrors of interest, but as aspirational engines that stimulate consumption through subtle nudging, iterative content exposure, and algorithmic hierarchies of visibility. On TikTok, the "For You" page delivers an endless scroll of short-form videos optimized by engagement-based machine learning models. Youth are frequently exposed to aestheticized consumption performances—fashion hauls, skincare routines, and product placements—typically enacted by influencers who conform to platform-specific ideals of attractiveness, charisma, and social popularity. Such content is algorithmically amplified according to metrics like watch time, likes, and interaction rates, effectively constructing desirability through visibility. As adolescents mimic popular formats or join viral challenges, their behaviors increasingly align with platform incentives rather than autonomous interest (Cotter 2019). ZEPETO similarly fosters consumption through its avatar customization economy. Features such as "recommended looks," trending items, and prominently displayed premium assets guide users toward purchasable virtual goods that embody dominant fashion norms. Algorithmically emphasized signals—e.g., badges, limited-time offers, or exclusive labels—create perceived scarcity and symbolic status, producing desire via social relevance cues. As one adolescent noted in a vlog, "I didn't even think about that bag until it kept showing up on my homepage. Suddenly, it became something I felt I had to have." This quote encapsulates how algorithmic repetition shapes not only awareness but emotional attachment and perceived necessity. Among youth negotiating identity, belonging, and peer status, these systems manufacture aspirational attachments to stylized goods and lifestyles. In this regard, platforms operate as cultural programmers, embedding consumption logics into everyday user experience. As Striphas(2015) and Bucher(2018) argue, algorithms serve not only as technical systems but as ideological infrastructures, organizing visibility in ways that discipline desire and regulate self-presentation.

5.2 Theme 2: Visibility Economy and Digital Peer Validation

Contemporary digital platforms increasingly function as visibility economies, where attention constitutes a central form of capital and algorithmic systems mediate access to social validation. Within these ecosystems, youth engagement is both expressive and strategically calibrated for exposure, as visibility becomes synonymous with influence, status, and even monetization (Abidin 2020; Marwick 2013). On TikTok, users commonly adopt platform-sanctioned behaviors—participating in trending challenges, using viral filters, syncing videos to popular audio tracks—not merely for creative expression but to increase algorithmic promotion. The pursuit of virality has transformed into a social imperative, tied to peer recognition, digital reputation, and identity affirmation. In this logic, visibility becomes value, and metrics such as likes and shares operate as proxies for cultural worth. Roblox mirrors this dynamic through its creator economy, where youth design and publish user-generated games. Games that reflect popular themes—luxury brands, influencer lifestyles, or celebrity culture—tend to attract greater engagement and earn more Robux (the platform's currency), particularly when aligned with dominant visual and narrative aesthetics. This reward system conditions youth creators to prioritize trend-compatible content over experimental or authentic expression, reinforcing a model in which algorithmic legibility determines success. The platform architecture embeds a logic of performative optimization, requiring youth to continuously manage their digital personas in accordance with anticipated audience reception and algorithmic criteria. This aligns with Marwick (2013)'s notion of context collapse, wherein the diverse audiences of digital life—peers, strangers, algorithms, and brands—are compressed into a single imagined public. In response, youth often cultivate generalized, broadly appealing identities that sacrifice complexity in favor of recognizability and reach.

5.3 Theme 3: Datafication and the Internalization of Market Norms

The third theme reveals how the datafication of youth activity fosters the internalization of market-oriented values within platform-mediated identity formation. Through the extraction, analysis, and monetization of behavioral data, platforms reframe participation as a commercial process, wherein every click, scroll, and customization is converted into a datapoint within a predictive system designed to optimize user behavior toward profitable outcomes (Van Dijck and Poell 2013; Zuboff 2019). In this context, youth are positioned not only as content consumers but also as producers of monetizable data. Their behaviors are continuously tracked and algorithmically modeled to shape future exposure, refine personalization, and sustain engagement. While these backend processes remain largely opaque to users, they profoundly structure how identity, aspiration, and social value are constructed and sustained. On ZEPETO, for example, the personalization engine recommends items, filters, and digital environments based on user interaction histories. This generates a data-driven feedback loop, reinforcing preferences that align with dominant commercial aesthetics—attractiveness, trendiness, and brand affiliation. Youth whose stylized choices receive more algorithmic attention are more likely to repeat them, thereby aligning their self-presentation with platform-rewarded templates.

As one Roblox user remarked in a forum post: "I felt like I had to keep upgrading my avatar to get noticed. Everyone with cooler stuff got more friends." This quote exemplifies how platform logic equates social belonging with consumption performance, linking visibility and connection to digital expenditures. In such environments, youth internalize the belief that success and recognition are contingent on conformity to aesthetic and market norms. These dynamics reflect what Van Dijck and Poell (2013) terms platformization—the restructuring of social practices around commercially-driven infrastructures. Here, identity commodification is not confined to explicit advertising but occurs through everyday micro-interactions: likes, purchases, customizations. As Couldry and Mejias (2019) notes, such interactions normalize a consumerist habitus, in which personal value is tethered to algorithmic recognition and market compatibility. Over time, this process cultivates a worldview in which consumer performance is equated with self-worth, and algorithmic approval becomes a proxy for social legitimacy, subtly shaping the values, desires, and digital identities of the younger generation.

6 Conclusion

This study has explored how big data and algorithmic infrastructures shape youth consumption culture within AI-driven virtual environments, using a theory-informed and interdisciplinary qualitative case study approach. By analyzing platforms such as TikTok, ZEPETO, and Roblox, the research demonstrates that these digital spaces function not merely as sources of entertainment but as socio-technical environments where youth identity, aspiration, and value systems are algorithmically curated and economically instrumentalized.

At the heart of this transformation lies a fundamental shift in the position of youth—from passive consumers to data-producing cultural agents. Their attention, behaviors, and self-representations are continuously collected, analyzed, and monetized. These algorithmic systems do not merely reflect pre-existing preferences; rather, they actively construct desire, regulate identity performances, and shape normative standards of social and cultural legitimacy. In this sense, youth consumption culture becomes deeply entangled with the logic of personalization, visibility, and market reproduction embedded in platform capitalism.

Theoretically, this study contributes to critical platform studies and digital youth research by integrating the political economy of datafication with theories of algorithmic culture and consumer socialization. It offers a critical lens to understand how platform infrastructures mediate and reshape cultural norms, values, and practices among younger generations. On a practical level, the findings provide important implications for educators, platform designers, and policymakers. The study emphasizes the urgency of implementing platform literacy education and developing youth-centered design frameworks that promote autonomy, diversity, and critical digital engagement.

However, the study also acknowledges several limitations. First, while TikTok, ZEPETO, and Roblox are globally influential, they do not encompass the full range of youth digital experiences, particularly across diverse cultural, linguistic, and geopolitical contexts. Second, the reliance on publicly accessible user-generated content may underrepresent private, marginalized, or less-visible perspectives. Third, the cross-sectional nature of the research provides a snapshot of platform engagement, rather than capturing long-term shifts in youth identity or consumption patterns.

Future research can address these limitations by conducting longitudinal studies that trace how youth interactions with algorithmic systems evolve over time, or by comparing cross-cultural dynamics to uncover variations in datafication processes across different social settings. Moreover, participatory methodologies involving youth as co-researchers may help amplify their lived experiences and resist the objectifying tendencies of data extraction. Beyond these scholarly contributions, this study also points to the need for policy and educational interventions. Policymakers should consider regulations that enforce transparency in data use, algorithmic accountability, and youth rights in digital environments. Educational institutions, in turn, must incorporate digital literacy programs that equip youth with the critical skills necessary to navigate, question, and challenge platform logics. As algorithmic systems continue to embed themselves in the everyday lives of adolescents, a deeper understanding of their cultural consequences is not only timely but essential for shaping equitable and ethical digital futures for younger generations.

References

Abidin C (2020) Mapping Internet celebrity on TikTok: exploring attention economies and visibility labours. Cult Sci J 12(1):77–103. https://doi.org/10.5334/csci.140

Bishop S (2021) Influencer industry: the quest for authenticity on social media. Polity Press

Boellstorff T, Nardi B, Pearce C, Taylor TL (2012) Ethnography and virtual worlds: a handbook of method. Princeton University Press

Braun V, Clarke V (2006) Using thematic analysis in psychology. Qual Res Psychol 3(2):77–101. https://doi.org/10.1191/1478088706qp063oa

Bucher T (2018) If...then: Algorithmic power and politics. Oxford University Press

Chan K (2013) Development of materialistic values among children and adolescents. Young Consum 14(3):244–256. https://doi.org/10.1108/YC-07-2013-00361

Cotter K (2019) Playing the visibility game: how digital influencers and algorithms negotiate influence on Instagram. New Media Soc 21(4):895–913. https://doi.org/10.1177/1461444818802974

Couldry N, Mejias UA (2019) The costs of connection: how data is colonizing human life and appropriating it for capitalism. Stanford University Press

de Reuver M, Sørensen C, Basole RC (2020) The digital platform: a research agenda. J Inf Technol 35(2):124–135. https://doi.org/10.1177/0268396220932033

Dittmar H (2008) Consumer culture, identity and well-being: the search for the 'good life' and the 'body perfect.' Psychology Press

Duffy BE (2017) (Not) getting paid to do what you love: gender, social media, and aspirational work. Yale University Press

Erikson EH (1968) Identity: youth and crisis. W. W. Norton & Company

Eubanks V (2018) Automating inequality: how high-tech tools profile, police, and punish the poor. Martin's Press, St

Flyvbjerg B (2006) Five misunderstandings about case-study research. Qual Inq 12(2):219–245. https://doi.org/10.1177/1077800405284363

Gillespie T (2014) The relevance of algorithms. In: Gillespie T, Boczkowski P, Foot K (eds) Media technologies: essays on communication, materiality, and society. MIT Press, pp 167–194

Hine C (2015) Ethnography for the internet: embedded, embodied and every day. Bloomsbury Academic

James A, Prout A (1997) Constructing and reconstructing childhood: contemporary issues in the sociological study of childhood, 2nd edn. Routledge

John DR (1999) Consumer socialization of children: a retrospective look at twenty-five years of research. J Consum Res 26(3):183–213. https://doi.org/10.1086/209559

Livingstone S (2008) Taking risky opportunities in youthful content creation: teenagers' use of social networking sites for intimacy, privacy and self-expression. New Media Soc 10(3):393–411. https://doi.org/10.1177/1461444808089415

Livingstone S, Blum-Ross A (2020) Parenting for a digital future: how hopes and fears about technology shape children's lives. Oxford University Press

Marwick A (2013) Status update: celebrity, publicity, and branding in the social media age. Yale University Press

Montgomery KC (2015) Youth and surveillance in the digital age: global trends and local realities. Springer

Nandagiri R, Ghaffar F (2022) Youth on TikTok: exploring algorithmic consumption and identity formation. J Youth Stud 25(9):1243–1261. https://doi.org/10.1080/13676261.2021.1927980

Noble SU (2018) Algorithms of oppression: how search engines reinforce racism. NYU Press.

Oh D, Lee Y (2021) Avatars and identity performance in immersive virtual environments: the case of ZEPETO. Games Cult 16(6):741–761. https://doi.org/10.1177/1555412021999844

Papacharissi Z (2015) Affective publics: sentiment, technology, and politics. Oxford University Press

Patton MQ (2015) Qualitative research and evaluation methods, 4th edn. SAGE Publications

Pink S, Horst H, Postill J, Hjorth L, Lewis T, Tacchi J (2016) Digital ethnography: principles and practice. SAGE Publications

Striphas T (2015) Algorithmic culture. Eur J Cult Stud 18(4–5):395–412. https://doi.org/10.1177/1367549415577392

Van Dijck J, Poell T (2013) Understanding social media logic. Media Commun 1(1):2–14. https://doi.org/10.17645/mac.v1i1.70

Yin RK (2018) Case study research and applications: design and methods, 6th edn. SAGE Publications

Zuboff S (2019) The age of surveillance capitalism: the fight for a human future at the new frontier of power. PublicAffairs

Coaching Leadership and Organizational Culture: The Roles of Self Leadership, Emotional Labor, Organizational Commitment, and Human Resource Development

Mijeong Kim and Sunghee Baek

1 Introduction

In today's organizational landscape, rapid change and growing uncertainty have underscored the importance of internal factors such as employee autonomy, psychological competence, and the quality of Organizational Culture (Schein 2010; Swanson and Holton 2001). These elements have emerged as critical determinants of organizational performance (Jung 2024). Among them, leadership that fosters employee development and commitment has become increasingly vital in shaping a constructive and sustainable Organizational Culture (Grant 2014; Goleman 2000). Accordingly, scholarly interest in the influence of leadership on Organizational Culture has steadily increased (Schein 1985; Heslin et al. 2006).

Coaching Leadership has gained attention as a leadership style that emphasizes empowerment, feedback, and growth rather than directive control (Heslin et al. 2006; Grant 2014). By encouraging self-directed behavior and competence development, coaching leaders are positioned to enhance employee motivation and foster alignment with the organization's values and goals (Tenschert et al. 2024). However, while the theoretical potential of Coaching Leadership in shaping Organizational Culture is well recognized, the specific mechanisms by which this influence occurs remain underexplored. In particular, empirical evidence on the mediating or moderating variables involved in this relationship is still limited (Kim 2024a, b; Park 2025).

M. Kim (✉)
Program in Project Management, Soongsil University, Seoul, Republic of Korea
e-mail: perfumejung@naver.com

S. Baek
Department of Project Management, Soongsil University, Seoul, Republic of Korea
e-mail: b0s0h073@naver.com

R. Lee et al. (eds.), *IT Service & Business Innovation*, Studies in Computational Intelligence 1252, https://doi.org/10.1007/978-3-032-14397-6_6

This study seeks to examine the pathways through which Coaching Leadership influences Organizational Culture. Specifically, the research investigates the mediating roles of Self-Leadership, Emotional Labor, and Organizational Commitment, as well as the moderating effect of Human Resource Development (HRD). By adopting a multidimensional analytical framework, this study aims to provide a more comprehensive understanding of how leadership interacts with internal dynamics to shape Organizational Culture. The findings are expected to offer both theoretical insights and practical implications for leadership development and HRD strategy formulation.

The specific objectives of this study are as follows. First, to analyze the impact of Coaching Leadership on employees' perception of Organizational Culture. Second, to examine the mediating effects of Self-Leadership, Emotional Labor, and Organizational Commitment in this relationship. Third, to assess the moderating effect of HRD between Coaching Leadership and Organizational Culture.

This paper is structured as follows. Section 2 reviews the theoretical background and related literature, leading to the development of research hypotheses. Section 3 outlines the research model and methodology. Section 4 presents the results of the empirical analysis. Finally, Sect. 5 discusses the findings, implications, limitations, and suggestions for future research.

2 Theoretical Background

2.1 Coaching Leadership

Coaching Leadership, introduced as one of the six leadership styles, focuses on employee growth and development (Goleman 2000). Coaching leaders promote employee advancement through mutual feedback and self-reflection (Heslin et al. 2006). Rather than relying on command-and-control, this leadership style is grounded in trust-based relationships, empathetic listening, and individualized support. Coaching Leadership fosters autonomy and motivation, positively influencing organizational performance (Grant 2014). More recently, Coaching Leadership has been shown to significantly enhance strengths use and job engagement (Jung and Jung 2024). Coaching leaders encourage learning and help employees find their own solutions, thereby reinforcing autonomy and psychological safety within the organization.

2.2 Organizational Culture

The foundational theory of Organizational Culture was proposed by Schein, who defined it as the shared basic assumptions and beliefs held by organizational members (Schein 1985). Depending on leadership, Organizational Culture can range from conservative to open and learning-oriented (Schein 2010). Organizational Culture in small and medium-sized enterprises has been found to directly affect job satisfaction and commitment (Jung 2024). A positive culture contributes to employee engagement, performance, and identity formation.

2.3 *Self-Leadership*

Self-Leadership refers to a set of strategies individuals use to motivate and guide their own behavior (Manz 1986). These strategies include behavior-focused strategies, natural reward strategies, and constructive thought patterns. Self-Leadership has been shown to enhance self-efficacy, creativity, and responsibility (Neck and Houghton 2006). Self-Leadership training positively affects leadership development, particularly when combined with mindfulness (Tenschert et al. 2024). In organizational contexts, Self-Leadership promotes goal orientation and self-regulation, which can positively shape Organizational Culture.

2.4 *Emotional Labor*

Emotional Labor refers to the process of managing emotions to meet organizational expectations (Hochschild 1983). It is categorized into surface acting, which involves modifying outward emotional expressions, and deep acting, which involves internal emotional adjustment. Excessive Emotional Labor can lead to burnout (Brotheridge and Lee 2003). Coaching Leadership has been suggested as a factor that can help mitigate Emotional Labor (Kim 2024a, b). The internalization of emotional norms affects the development of Organizational Culture, and a healthy emotional climate can lead to a culture of mutual respect.

2.5 *Organizational Commitment*

Organizational Commitment was initially conceptualized as a construct by Porter et al. (1974), Mowday et al. (1979) and later refined into three dimensions—affective, continuance, and normative commitment—by Meyer and Allen (1991). Affective commitment reflects emotional attachment, continuance commitment involves awareness of the costs associated with leaving, and normative commitment stems from a sense of obligation. High levels of commitment are associated with reduced turnover, enhanced organizational citizenship behavior, and improved performance. Self-efficacy and job satisfaction have been shown to influence Organizational Commitment (Kim 2025a, b), while Coaching Leadership has been found to indirectly enhance customer orientation through the mediating role of commitment (Park

2025). Commitment plays a key role in integrating employees into Organizational Culture and internalizing organizational values.

2.6 *Human Resource Development (HRD)*

Human Resource Development (HRD) was first theorized as a formal concept in the early 1970s (Nadler 1970) and was later elaborated as a function of organizational learning, performance, and change (Swanson and Holton 2001). HRD encompasses training, career development, and organizational development activities aimed at enhancing individual capabilities and job performance. It has been shown to positively influence Organizational Commitment as well as internal and external competitiveness (Kim 2024a, b). The impact of HRD on job satisfaction and professional expertise has also been empirically validated, particularly in the public sector (Jin 2025). Within organizations, HRD can interact with Coaching Leadership to strengthen employee commitment and foster a supportive Organizational Culture.

3 Research Model and Methodology

3.1 *Research Model*

3.1.1 Relationship Between Coaching Leadership and Organizational Culture

Coaching Leadership, characterized by its emphasis on employee development and mutual interaction rather than traditional directive control, plays a vital role in building trust and enhancing motivation (Heslin et al. 2006). By fostering psychological safety, learning orientation, and autonomy within organizations, Coaching Leadership contributes to the formation of a constructive and innovative Organizational Culture (Grant 2014). Organizational Culture, defined as the shared values, beliefs, and behavioral norms among members, is a key determinant of organizational sustainability and innovation (Schein 2010). Leaders act as facilitators of culture, and coaching leaders, in particular, are well-positioned to enhance cultural quality by encouraging upward communication and aligning leadership behaviors with employees' intrinsic motivations and values.

H1. Coaching Leadership has a positive effect on employees' perception of Organizational Culture.

3.1.2 Relationship Between Coaching Leadership and Self-Leadership, Emotional Labor, and Organizational Commitment

Coaching Leadership has been found to be an important antecedent of Self-Leadership, as it strengthens employee autonomy and responsibility (Manz 1986; Grant 2014). By offering constructive feedback and support, coaching leaders enable employees to develop self-regulatory and self-motivational strategies (Tenschert et al. 2024). Moreover, Coaching Leadership can mitigate the negative effects of Emotional Labor by encouraging deep acting over surface acting, thus helping employees maintain emotional balance and job satisfaction (Hochschild 1983; Kim 2024a, b). Employees who experience Coaching Leadership are more likely to regulate emotional expression effectively, contributing to a positive work climate. In addition, Coaching Leadership fosters emotional bonds with employees and strengthens their attachment to the organization (Allen and Meyer 1991; Park 2025), which in turn enhances Organizational Commitment.

H2. Coaching Leadership has a positive effect on employees' Self-Leadership.

H3. Coaching Leadership has a positive effect on employees' Emotional Labor.

H4. Coaching Leadership has a positive effect on employees' Organizational Commitment.

3.1.3 Relationship Between Self-Leadership, Emotional Labor, Organizational Commitment, and Organizational Culture

Employees with high levels of Self-Leadership tend to demonstrate greater understanding and engagement with organizational goals and values, thereby promoting internalization and dissemination of culture (Manz 1986; Jung 2024). Their autonomous behavior also positively influences coworkers, fostering a community-oriented culture. Furthermore, when Emotional Labor is performed primarily through deep acting rather than surface acting, it contributes to higher-quality Organizational Culture (Kim et al. 2025a, b). The internalization of emotional norms strengthens mutual respect and psychological stability within the organization. Organizational Commitment, defined as the psychological attachment and desire to remain within an organization (Meyer and Allen 1991), is also a critical driver of cultural formation. Committed employees internalize organizational values and serve as role models for others (Kim 2025a, b).

H5. Employees' Self-Leadership has a positive effect on Organizational Culture.

H6. Employees' Emotional Labor has a positive effect on Organizational Culture.

H7. Employees' Organizational Commitment has a positive effect on Organizational Culture.

3.1.4 Mediating Roles of Self-Leadership, Emotional Labor, and Organizational Commitment

Self-Leadership refers to the ability of individuals to set goals, self-motivate, and regulate their behavior (Manz 1986). Coaching Leadership supports this self-directed behavior through feedback and reflective dialogue, which ultimately enhances cultural internalization and proactive behavior within the organization. Emotional Labor, the act of regulating or expressing emotions according to organizational norms (Hochschild 1983), can cause job stress and burnout. Coaching leaders can buffer these negative effects through emotional support and empathy, creating a healthier emotional environment that fosters positive cultural perception. Organizational Commitment, which reflects employees' emotional attachment and willingness to remain in the organization (Meyer and Allen 1991), serves as a key mechanism in the internalization of values and engagement in cultural behaviors. By focusing on individual growth and needs, Coaching Leadership is expected to enhance commitment, thereby strengthening Organizational Culture.

H8. Self-Leadership mediates the relationship between Coaching Leadership and Organizational Culture.

H9. Emotional Labor mediates the relationship between Coaching Leadership and Organizational Culture.

H10. Organizational Commitment mediates the relationship between Coaching Leadership and Organizational Culture.

3.1.5 Moderating Effect of Human Resource Development (HRD)

HRD encompasses training, education, and development systems aimed at enhancing employee competencies and supporting continuous growth (Swanson and Holton 2001). The developmental orientation of Coaching Leadership is expected to interact positively with the organizational HRD environment. Depending on the level of HRD support, the effectiveness of Coaching Leadership may vary. Accordingly, this study assumes that HRD moderates the effect of Coaching Leadership on Organizational Culture.

H11. Human Resource Development moderates the relationship between Coaching Leadership and Organizational Culture.

Based on the above theoretical discussion, this study proposes a total of eleven hypotheses: H1–H4 reflect the direct effects, H5–H10 reflect the mediating effects, and H11 represents the moderating effect. The overall research model is illustrated in Fig. 1.

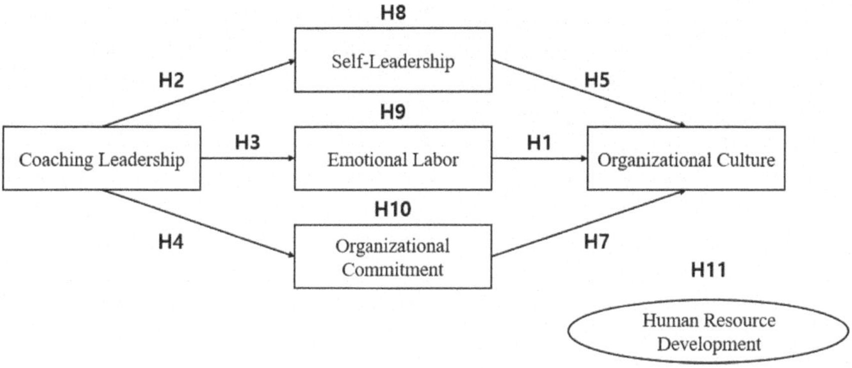

Fig. 1 The research model

3.2 Sample and Data Collection

This study targeted employees working in large corporations and public organizations who had prior experience with coaching or leadership-related training. Data were collected through a structured online survey conducted over a one-week period from May 12 to May 18, 2025. Participants were informed of the study's purpose and provided voluntary consent prior to participation. A total of 263 responses were initially collected, and after excluding 5 cases due to insincere or incomplete responses, 253 valid responses were retained for analysis.

3.3 Definition and Measurement of Variables

The variables in this study were defined and measured based on established instruments from prior research. All items were rated on a five-point Likert scale (1 = Strongly Disagree, 5 = Strongly Agree). A summary of each variable, including its source and conceptual focus.

3.4 Analytical Strategy

To empirically test the proposed hypotheses, data analysis was conducted using SPSS 25.0 and PROCESS Macro 4.2 for SPSS (Hayes 2022). The analytical procedure was carried out in several stages to ensure statistical rigor and clarity in interpreting the results.

First, descriptive statistics—including means, standard deviations, skewness, kurtosis, and correlation coefficients—were calculated to examine the basic characteristics and relationships among the study variables. This preliminary step provided an overview of the data distribution and inter-variable associations.

Next, reliability analysis was conducted to verify the internal consistency of each measurement scale. Cronbach's alpha coefficients were computed to assess the degree to which the items within each construct measured the same underlying concept.

To examine the direct effects of Coaching Leadership on Self-Leadership, Emotional Labor, Organizational Commitment, and Organizational Culture (H1–H4), simple linear regression analyses were performed. These tests aimed to confirm the foundational relationships in the proposed model.

Following this, the mediating roles of Self-Leadership, Emotional Labor, and Organizational Commitment (H5–H10) were tested using bootstrapping techniques via Model 4 of the PROCESS Macro. 95% confidence intervals were calculated to assess the significance of indirect effects.

Lastly, to assess the moderating effect of Human Resource Development (HRD) on the relationship between Coaching Leadership and Organizational Culture (H11), Model 1 of the PROCESS Macro was applied. In addition to the interaction term's significance, simple slope analysis was conducted to interpret the moderation pattern in detail.

All analyses were conducted at the 0.05 significance level. Hypotheses were evaluated based on the direction and statistical significance of the regression coefficients.

4 Results

4.1 Confirmatory Factor Analysis

To evaluate the validity of the measurement model, a confirmatory factor analysis (CFA) was conducted. Each construct in the study was modeled as a single-factor structure. For constructs with a large number of items, an item parceling technique was applied, grouping items into three parcels to improve model parsimony and stability.

The results of the CFA indicated a good model fit: $\chi^2/df = 1.95$, CFI = 0.962, TLI = 0.947, and RMSEA = 0.048. In addition, all factor loadings (λ) exceeded the recommended threshold of 0.70, average variance extracted (AVE) values were above 0.50, and construct reliability (CR) values exceeded 0.70. These results confirm that the measurement model satisfies the criteria for convergent validity and internal consistency.

Table 1 Means, standard deviations, and correlations among variables

Variable	M	SD	1	2	3	4	5	6
Coaching leadership	3.86	0.64	(0.910)					
Self-leadership	3.78	0.49	0.389**	(0.828)				
Emotional labor	3.94	0.62	0.366**	0.646**	(0.886)			
Organizational commitment	3.99	0.65	0.650**	0.317**	0.348**	(0.892)		
Organizational culture	3.92	0.54	0.540**	0.681**	0.732**	0.620**	(0.911)	
Human resource development	3.98	0.48	0.423**	0.541**	0.413**	0.315**	0.525**	(0.858)

Note ** $p < 0.01$, numbers in parentheses are Cronbach's alphas

4.2 Descriptive Statistics and Correlation Analysis

Before conducting hypothesis testing, descriptive statistics and inter-variable correlations were examined. The means and standard deviations of the main variables were within acceptable ranges, and the data satisfied the assumption of normality. All correlations were in the expected theoretical directions and showed statistically significant positive relationships. Additionally, the internal consistency of each scale was confirmed through reliability analysis, with Cronbach's alpha coefficients exceeding 0.80 for all constructs, indicating acceptable reliability (Table 1).

4.3 Hypothesis Testing

4.3.1 Direct Effects (H1–H4)

To examine the direct effects of Coaching Leadership on key outcome variables, simple linear regression analyses were conducted. The results demonstrated that Coaching Leadership significantly and positively influenced all four dependent variables in the model. Specifically, it had a strong positive impact on organizational culture, supporting Hypothesis 1. Coaching leader-ship also significantly predicted Self-Leadership, Emotional Labor, and Organizational Commitment, providing support for Hypotheses 2 through 4. These findings confirm that Coaching Leadership plays a foundational role in shaping both psychological and organizational outcomes among employees (Table 2).

Table 2 Results of simple regression analyses (H1–H4)

Dependent variable	B	SE	β	t	p	R^2	F
Organizational culture	0.456	0.044	0.540	10.271***	0.000	0.289	105.494
Self-leadership	0.296	0.044	0.389	6.766***	0.000	0.148	45.777
Emotional labor	0.354	0.056	0.366	6.291***	0.000	0.131	39.575
Human resource development	0.661	0.048	0.650	13.672***	0.000	0.420	186.927

Note *** $p < 0.001$

4.3.2 Indirect Effects of Mediators (H5–H10)

The mediating effects were tested using Model 4 of the PROCESS Macro for SPSS (Hayes 2022), applying a bootstrapping method with 5,000 resamples. In this analysis, Coaching Leadership was specified as the independent variable (X), Organizational Culture as the dependent variable (Y), and three mediators were tested individually: Self-Leadership (M1), Emotional Labor (M2), and Organizational Commitment (M3). The significance of indirect effects was evaluated based on the 95% bootstrap confidence intervals (Table 3).

The results of the bootstrapping analysis confirmed that all three mediating effects were statistically significant. Specifically, the indirect effect of Coaching Leadership on Organizational Culture through Self-Leadership was significant (indirect effect = 0.183, SE = 0.032, 95% CI [0.1199, 0.2468]). Similarly, Emotional Labor (indirect effect = 0.191, SE = 0.036, 95% CI [0.1219, 0.2609]) and Organizational Commitment (indirect effect = 0.255, SE = 0.042, 95% CI [0.1744, 0.3395]) also mediated the relationship significantly. Since all confidence intervals excluded zero, the mediating effects of the three variables were supported (Table 4).

4.3.3 Moderation Analysis Using PROCESS Macro (H11)

Finally, the moderating effect of Human Resource Development (HRD) on the relationship between Coaching Leadership and Organizational Culture was examined. Model 1 of the PROCESS Macro (Hayes 2022) was used, incorporating the interaction term between Coaching Leadership and HRD.

Coaching Leadership had a significant positive effect on Organizational Culture ($B = 0.315$, $p < 0.001$), and HRD also positively predicted Organizational Culture ($B = 0.433$, $p < 0.001$). Importantly, the interaction term (Coaching Leadership × HRD) was found to be statistically significant ($B = 0.198$, $p < 0.01$), and the change in R^2 after adding the interaction term was 0.017, which was also significant ($p < 0.01$). These results support Hypothesis 11, confirming that HRD moderates the relationship between Coaching Leadership and Organizational Culture (Table 5).

Table 3 Results of mediation analysis using process macro (model 4)

Path	β	SE	t	p	95% CI	
					LLCI	ULCI
X → M1 (self-leadership)	0.389	0.043	6.766***	0.000	0.2099	0.3823
M1 → Y (organizational culture)	0.616	0.051	12.216***	0.000	0.5173	0.7162
X → Y (direct effects)	0.274	0.044	10.271***	0.000	0.1981	0.3493
X → M2 (emotional labor)	0.353	0.056	6.291***	0.000	0.2430	0.4644
M2 → Y (organizational culture)	0.539	0.036	14.912***	0.000	0.1967	0.3344
X → Y (direct effects)	0.266	0.035	7.596***	0.000	0.4682	0.6106
X → M3 (organizational commitment)	0.661	0.048	13.672***	0.000	0.5656	0.7559
M3 → Y (organizational culture)	0.386	0.052	7.393***	0.000	0.2834	0.4891
X → Y (direct effects)	0.201	0.053	3.785***	0.000	0.0965	0.3058
X → Y (total effects)	0.456	0.038	7.123***	0.000	0.3689	0.5439

Note *** $p < 0.001$, Independent Variable: Coaching Leadership, dependent Variable: Organizational Culture

Table 4 Bootstrapped indirect effects for mediating paths (model 4)

ResultPath	Indirect Effect	SE	95% CI		Significance
			LLCI	ULCI	
X → M1 → Y	0.183	0.032	0.1199	0.2468	Significance
X → M2 → Y	0.191	0.036	0.1219	0.2609	Significance
X → M3 → Y	0.255	0.042	0.1744	0.3395	Significance

Note *** $p < 0.001$, Independent Variable: Coaching Leadership, dependent Variable: Organizational Culture

The interaction term between Coaching Leadership and Human Resource Development (HRD) was found to be statistically significant. To further interpret this moderating effect, a conditional effects analysis was conducted at three levels of HRD: the mean (M), one standard deviation below the mean (M − 1SD), and one

Table 5 Results of moderation analysis using process macro (model 1)

Variable		B	SE	t	p	95% CI	
						LLCI	ULCI
Constant		3.891	0.028				
Independent variable	Coaching leadership	0.315	0.045	6.997***	0.000	0.2260	0.4031
Moderator	Human resource development	0.433	0.060	7.214***	0.000	0.3151	0.5517
Interaction term	Coaching leadership	0.198	0.073	2.724**	0.007	0.0547	0.3405
	Human resource development						
ΔR^2 due to interaction		R^2		F		p	
		0.017		7.417		0.007	

Note *** p < 0.001, dependent Variable: Organizational Culture

Table 6 Conditional effects by level of HRD (moderation analysis)

Conditional indirect effects	Effect	SE	t	p	95% CI	
					LLCI	ULCI
M − 1SD	0.205	0.063	3.236	0.001	0.0803	0.3300
M (0.000)	0.318	0.445	7.093	0.000	0.2297	0.4063
M + 1SD	0.403	0.053	7.667	0.000	0.2993	0.5062

standard deviation above the mean (M + 1SD). The results indicated that the effect of Coaching Leadership on Organizational Culture was significant across all levels of HRD, with stronger effects observed as HRD increased. These findings support the presence of a moderating effect of HRD, confirming that the influence of Coaching Leadership on Organizational Culture varies depending on the level of HRD (Table 6).

5 Conclusion

This study empirically investigated the impact of Coaching Leadership on Organizational Culture, focusing on the mediating role of Self-Leadership, Emotional Labor, Organizational Commitment, and Human Resource Development (HRD). The positive effect of Coaching Leadership on performance and autonomous behavior was demonstrated, and the significant mediating effect of Self-Leadership, Emotional Labor, and Organizational Commitment was further confirmed through bootstrapping analysis using PROCESS macro model 4. In addition, the moderating role of HRD indicates that HRD acts as a contextual enhancer of Coaching Leadership due to the significant impact of the interaction between Coaching Leadership and HRD on Organizational Culture.

Theoretical Implications: This study contributes to the literature by presenting an integrated multivariate model that explains the pathways through which Coaching Leadership shapes Organizational Culture. By simultaneously considering cognitive and emotional mediators—namely Self-Leadership, Emotional Labor, and Organizational Commitment—the study provides a deeper understanding of the mechanisms underlying cultural formation. Furthermore, confirming the moderating role of HRD demonstrates a meaningful theoretical contribution by highlighting the interaction between leadership and human resource systems (Swanson and Holton 2001).

Practical Implications: Organizations aiming to cultivate a high-performing culture should consider adopting leadership development programs that emphasize coaching-based approaches. Structured feedback mechanisms can support leaders in promoting autonomy, fostering development, and aligning team behavior with organizational values. At the employee level, emotional intelligence training and peer coaching initiatives may serve as effective tools to enhance Self-Leadership while reducing the psychological strain associated with Emotional Labor. These interventions can strengthen emotional resilience and deepen employee engagement. Moreover, investing in comprehensive Human Resource Development systems including training infrastructure and performance evaluation frameworks will amplify the positive outcomes of Coaching Leadership. A strategically designed HRD environment can ensure that leadership development efforts are not only sustained but also embedded across all levels of the organization.

References

Allen NJ, Meyer JP (1991) A three-component conceptualization of organizational commitment. Hum Resour Manag Rev 1(1):61–89

Brotheridge CM, Lee RT (2003) Development and validation of the emotional labour scale. J Occup Organ Psychol 76(3):365–379

Goleman D (2000) Leadership that gets results. Harv Bus Rev. Accessed 24 June 2016

Grant AM (2014) Autonomy support, relationship satisfaction and goal focus in the coach–coachee relationship: which best predicts coaching success? Coach: Int J Theory Res Pract 7(1):18–38

Heslin PA, Vandewalle D, Latham GP (2006) Keen to help? Managers' implicit person theories and their subsequent employee coaching. Pers Psychol 59(4):871–902

Hochschild AR (1983) The managed heart: commercialization of human feeling. University of California Press

Jin SM (2025) A study on the role of civil servant training institutions for enhancing public officials' professionalism. Doctoral dissertation, Graduate School of Public Administration, Korea University, Seoul, South Korea

Jung SC (2024) The effect of organizational culture on job satisfaction and organizational commitment of small and medium-sized manufacturing companies: focusing on the moderating effect of organizational trust. J Internet and E-Commer 24(1):205–223

Jung HG, Jung JJ (2024) A study on the effect of coaching leadership on job enthusiasm: focusing on the mediating effect of strengths utilization and the moderating effect of leader-member exchange relationship. J Korean Manag Assoc 37(10):1731–1751
Kim IH (2024a) The effects of supervisors' coaching leadership on department store employees' emotional labor. Chung-Ang University
Kim JH (2024b) The effect of HRD intention and implementation level on internal and external competitiveness of firms: mediating role of organizational commitment. J Converg Cult Technol 10(3):259–265
Kim HY (2025a) Relationships among self-efficacy, job satisfaction, organizational commitment, and job performance. Seokyeong University
Kim SJ (2025b) Effects of supervisors' coaching leadership on organizational commitment. Korea University of Technology and Education
Manz CC (1986) Self-leadership: toward an expanded theory of self-influence processes in organizations. Acad Manag Rev 11(3):585–600
Meyer JP, Allen NJ (1991) A three-component model conceptualization of organizational commitment. Hum Resour Manag Rev 1(1):61–89
Mowday RT, Steers RM, Porter LW (1979) The measurement of organizational commitment. J Vocat Behav 14(2):224–247
Nadler L (1970) Developing human resources. Gulf Publishing Company
Neck CP, Houghton JD (2006) Two decades of self-leadership theory and research: past developments, present trends, and future possibilities. J Manag Psychol 21(4):270–295
Park SM (2025) Structural relationships among coaching leadership, self-efficacy, organizational commitment, and customer orientation. Hanyang University
Porter LW, Steers RM, Mowday RT, Boulian PV (1974) Organizational commitment, job satisfaction, and turnover among psychiatric technicians. J Appl Psychol 59(5):603–609
Schein EH (1985) Defining organizational culture. Class Organ Theory 3(1):490–502
Schein EH (2010) Organizational culture and leadership, 4th edn. Jossey-Bass
Swanson RA, Holton EF (2001) Foundations of human resource development. Berrett-Koehler
Tenschert J, Furtner M, Peters M (2024) The effects of self-leadership and mindfulness training on leadership development: a systematic review. Manag Rev Q 1–52

The Effect of Emotion Coaching Activities on Self-Regulation in 5 Year-Old Children

Sung-Hee Baek, Ho-Chul Shin, and Cheong-Ha Yeum

1 Introduction

In modern society, as the role of the family weakens due to smaller family units and an increase in dual-income households, Children's face challenges in acquiring the necessary skills for healthy interpersonal relationships and receiving adequate emotional training within the family. This environmental shift they are experiencing limits opportunities for social learning within the home and, coupled with an educational emphasis on cognitive development, has gradually reduced their social interaction opportunities (Eunyoung 2007). Consequently, contemporary preschoolers, despite exhibiting higher intellectual levels, tend to show a lack of patience, difficulty in managing their emotions, and a reduced ability to recognize or consider others' emotions (Jihyeon 2016). These issues come from a deficiency in self-regulation, the ability to monitor and control oneself, highlighting the necessity for effective education to address these problems (Mihwa 2013).

Self-regulation, in contrast to external control, is defined as the emotional capacity of children's to independently adjust their thoughts and feelings according to the situation without adults' instructions or demands (Yujin 2010). From a cognitive perspective, self-regulation encompasses higher-order cognitive abilities such as planning, monitoring, evaluating, and adjusting one's behavior (Jeongran 2003), and can be considered as the ability to regulate one's emotions and behavior within an acceptable range for successful social adaptation. Early childhood is recognized as a fundamental period for the development of self-regulation (Okseung 2006). An children's self-regulation does not develop innately but rather through socialization processes,

S.-H. Baek · C.-H. Yeum
Department of Project Management, Graduate School of Soongsil University, Seoul, Korea
e-mail: b0s0h073@naver.com

H.-C. Shin (✉)
Department of Business Administration, Soongsil University, Seoul, Korea
e-mail: hshin@ssu.ac.kr

R. Lee et al. (eds.), *IT Service & Business Innovation*, Studies in Computational Intelligence 1252, https://doi.org/10.1007/978-3-032-14397-6_7

alongside the development of behavior, language, cognition, and emotions within a social context (Jeongran 2003). Particularly, 5 year-old children's are in a developmental stage characterized by impulsivity and a lack of patience (Eunju 2005), yet they also begin to demonstrate self-regulatory behavior and emotions as their physical and language skills advance in the socialization process (Eunjoo 2010). Therefore, it is anticipated that social interactions with teachers and peers in early childhood education settings during this period will advance Children's self-regulation as they learn to manage negative emotions and inhibit inappropriate behavior.

Emotional coaching refers to guiding children's to develop the ability to actively and smoothly deal with problems that may arise in relation to their emotions (Gettman and Gelfand 2007). Gottman et al. (2011), who scientifically experimented with, verified, and systemized emotion coaching, emphasized that children's should be respected as individuals, and teachers should understand their emotions as they are, empathize with them, and guide their emotions and behavior in a desirable direction. He also stated that emotional coaching should empower children's to develop the ability to cope appropriately and wisely on their own, rather than simply following teachers' instructions in resolving their emotions (Jihyeon 2016). From this perspective, emotional coaching, which focuses on children's ability to recognize and regulate their own emotions, can be considered to influence their self-regulation.

In the process of emotional coaching, teachers should read children's emotions, build trust, intimacy, and rapport, and instead of using words of blame, contempt, defensiveness, command, or reprimand, lead children to resolve their emotions independently through listening, accepting, and understanding their inner feelings (Grand 2003). Through teachers' emotional coaching, children can learn to regulate their own emotions, monitor and control their behavior, and solve problems in a desirable way (Seongae and Byeok 2012). Experiences of recognizing, accepting, and resolving their own and others' emotions at their developmental level can play an important role in the development of self-regulation in early childhood (Hyonkyong 2012). Accordingly, the approach that utilizes emotional coaching to support children's self-regulation is considered meaningful.

Therefore, this study aims to explore the effect of providing emotion coaching activities on the self-regulation of 5 year-old children's, a developmental stage when they begin to have active emotional expressions and control their own thoughts and behavior. Especially, this study seeks to go beyond examining the effectiveness at the construct level of self-regulation and more specifically verify the effectiveness at the sub-factor level. By doing so, it aims to confirm the educational value of emotional coaching in early childhood and to provide fundamental data for developing emotional coaching programs to enhance children's self-regulation. In addition, to specifically examine the influence of emotional coaching programs utilizing personality activities on the self-regulation of 5 year-olds, research questions, divided into self-regulation and self-control, are established as the following:

Research Question 1. What is the effect of emotional coaching programs utilizing personality activities on 5 year-olds' self-regulation?

1-1. What is the effect of emotional coaching programs utilizing personality activities on self-evaluation, the sub-factor of self-regulation?
1-2. What is the effect of emotional coaching programs utilizing personality activities on self-determination, the sub-factor of self-regulation?

Research Question 2. What is the effect of emotional coaching programs utilizing personality activities on 5-year-olds' self-control?

2-1. What is the effect of emotional coaching programs utilizing personality activities on behavioral inhibition, the sub-factor of self-control?
2-2. What is the effect of emotional coaching programs utilizing personality activities on emotionality, the sub-factor of self-regulation?

2 Theoretical Background

2.1 Children' Self-Regulation

Self-regulation is related to the ability to recognize and control one's own emotions (Seongeui and Mihyang 2019). An infant with an experience of positive emotions develop their ability to understand and express their emotional state through self-regulation (Jihyeon 2010). For example, children who have experienced a happy moment can recognize their feelings as positive emotions and tend to share them with others (Hyesoon 2011). This process plays an important role in building close relationships with people and things (Jihye 2015).

Negative emotions play a role of inducing children to recognize something wrong and seek a solution to it (Misook 2016). For instance, an children's with an experience of anxiety could understand how they feel and take an action to get over it (Yooni 2013) In the process, the children's learns to display and control their emotions and gradually develops the power of adaptation (Seongeun 2013) In addition, children's with a high level of self-control don't act on impulse, even when they experience negative emotions, rather they have the tendency to analyze their emotions and take a proper action (Yeonju 2013). On the other hand, children's with a low level of self-regulation, under the strong influence of negative emotions, have difficulty in controlling their emotions and it can have a negative impact on building social relationships (Hyesoon 2011).

Consequently, to foster the ability to control negative emotions accordingly, an educational approach utilizing self-regulation is required (Eunju 2005) 'Social Relationship Part of the Revised Nuri Curriculum of 2019' also highlights children's self-regulation and emotional control and it includes descriptions such as 'exhibit my desire and emotion' and 'get interested in others' feelings and behavior. It suggests that emotional expression and regulation are closely connected to the formation of children's social relationships. Considering overall precedent studies, many of emotional education programs using picture books can be found, but emotional education programs along with personality education or emotional coaching hardly can be seen. Thus, this study seeks to develop an emotional education program using personality activities and emotional coaching, and, by introducing emotion plays to children, intends to combine them with the program.

3 Research Methods

3.1 Research Subjects

For this study, 30 five-year-old children's attending Kindergarten A, located in City P, Gyeonggi Province, were selected as the experimental group. Additionally, 30 five-year-old children enrolled in Kindergarten S, located in the same city, were selected as the comparison group. To control potential differences between two groups that could affect the study, both of kindergartens were situated in densely populated residential areas. Furthermore, the homeroom teachers of both groups had 7 years of teaching experience and majored in early childhood education. The children's in both groups were of the same age.

3.2 Measurement Tools

In this study, the tool used to examine the effect of emotional coaching on the self-regulation skills of five-year-old children is a self-regulation ability test. This tool was adapted for teachers based on Jeongran (2003) parent-report measure. The self-regulation ability test is not based on direct observation and evaluation of children within a fixed short time frame. Instead, homeroom teachers evaluate the developmental level of children's self-regulation skills during their daily routines. The self-regulation ability test is a measurement of self-regulation skills within the kindergarten classroom. The components of the self-regulation tool are 'self-monitoring' and 'self-control'. The sub-factors of self-monitoring are 'self-evaluation' and 'self-determination', while the sub-factors of self-control are 'behavioral inhibition' and 'emotionality'. The test consists of a total of 31 items, using a 5-point rating scale ranging from 'Not at all' (1 point) to 'Very much so' (5 points). The scale is directed so that a higher score, ranging from negative responses (1 point) to positive responses (5 points), indicates a higher level of self-regulation.

3.3 Research Procedure

The research procedure of this study consists of the following steps: development of emotional coaching activities, pre-test, teacher training, experimental treatment,

Table 1 Research design

Group	Pre-test	Treatment	Post-test
Experimental group	A1	X	A2
Comparison group	B1	O2	B2

and post-test. This study employed a quasi-experimental design, specifically a non-equivalent control group pre-test-post-test design. To examine the effect of emotional coaching on the self-regulation skills of five-year-old children, the self-regulation skills of the experimental group (A1) and the comparison group (B1) before the experimental treatment (X), were compared with the self-regulation skills of the experimental group (A1) and the comparison group (B1) after the experimental treatment (O). The specific research design is as shown in Table 1.

3.3.1 Pre-test

Prior to this study, a preliminary test was conducted for five days, to examine the adequacy of the program, the time span of activities, and the validity of the test tool. 12 five-year-old children's (5 boys, 7 girls) participated in the preliminary test, and they were of the same age as the experimental group but in a different class. The preliminary test aimed to investigate the interest level of the children, the adequacy of the activities, and reaction extent. Accordingly, the test was supplemented and revised. For an objective and precise evaluation, sufficient explanations were delivered to the homeroom teachers, and a checklist of measures to estimate the self-regulation ability was employed.

Before the pre-test, teacher training was conducted for two days for the teachers of the experimental group and the comparison group so that they sufficiently understand the test tool of this study and apply it to the program. The teacher of the experimental group was given a training about aims of educational activities, development of activities, emotional coaching methods, necessary materials, inquiry styles, cautions for instructions, which was implemented by two professional emotion coaches. During the program, continuous consultations were provided as well. Simulation classes were held to discuss on teaching methods and ensure that activities run smoothly.

After the researchers and the teachers of the experimental group and the comparison group were well informed of the rating method of the test tool, the pre-test was conducted, prior to the experimental treatment. And the pre-test employed a checklist for rating scales of self-regulation ability, based on the observations of the children's speech and behavior related to self-regulation skills during their daily routines.

3.3.2 Development of Activities

Emotional coaching is organized so that children's can recognize and display a variety of emotions that they can experience in everyday life and understand each other's emotions in their relationships with peers. In emotional coaching, a teacher plays a role of helping children's handle their own emotions properly and solve their problems with confidence, by encouraging them to understand their own emotions and learn how to solve their emotion-related problems as well as by empathizing with them and guiding them in the right direction. It stresses that a teacher should help children open their mind and encourage them to solve their problems in a desirable way, not making them follow teacher's instructions Thus, the teacher in the experimental group interacted with children, by saying things like "Oh, I see" and "It must be tough". And the teacher used emotional coaching methods and empathized with the children's when they talked about their own emotions, by saying things like "Oh, you are angry", "you feel terrible and upset". Five-year-old children are at the developmental stage when they begin to be self-assertive in interaction with peers and try to do on their own (Jeongran 2003). In addition, they begin to control themselves and follow others' requirements (Seongeun 2013). Being based on those behavior characteristics of five-year-old children, interactions between teachers and children, combined with emotional coaching methods, were applied. And the model for developing emotional coaching activities, based on emotional coaching theory of Gottman et al. (2011), Seongae and Byeok (2012), was organized as the following: 'Recognizing Emotions', 'Empathizing and Listening', 'Expressing emotions', and 'Resolving emotions' (Table 2).

The emotional coaching activities for this study were composed so that five-year-old children's independently recognize and display a variety of emotions that they could experience in everyday life and so that they understand each other's emotions in relationships with peers. The details of 10-session emotional coaching activities were designed after going through examinations by two early childhood educators and an in-service teacher. To be more specific, they were organized with applying desirable interactions between teachers and children's, based on activities with reference to studies by Seongae (2012) and Younghee (2014) and the emotional coaching theory of Gottman et al. (2011). Additionally, two early childhood educators and an in-service teacher helped to examine the process and the activities of emotional coaching. The

Table 2 Emotional coaching interaction

Emotional coaching	Teachers' interactions
Recognizing emotions	• You look angry, are you so? • How do you feel now?
Discovering emotions	• If you were ________, how would you feel? • If you were in this situation, how would you feel?
Empathizing	• So you are upset now—Oh, you are angry • Yeah, I have felt the same before
Expressing emotions	• Can you describe it with a drawing? • Can you tell me how you felt at the moment?
Resolving emotions	• What can you do when you have an argument with your friend? • What should you do in that case?

activities of the experimental group were held from Monday to Thursday every week, and the estimation on them was made by the teacher every Friday. To verify the effect of this study, after implementing five-week experimental treatments, a post-test was conducted on the experimental group and the comparison group over five days.

4 Analysis

The research was analyzed according to research questions with a view to exploring the impact of emotional coaching activities on children's self-regulation ability and the collected data were analyzed by using SPSS 18.0 program. As analysis techniques for the research, first, an independent sample t-test was conducted to check homogeneity of the experimental group and the comparison group, and Cronbach's α was implemented to examine the reliability of the measurement tool. Secondly, the independent sample t-test, namely difference verification between the experimental group and the comparison group was executed in three stages to investigate the influence of emotional coaching activities on five-year-old children's self-regulation. Stage 1 aims to explore the impact of emotional coaching strategy on five-year-old children's self-regulation. Stage 2 intends to examine the influence of emotional coaching strategy on the sub-factors of self-regulation, namely self-monitoring and self-control. Finally, stage 3 is set to specifically verify the effect of emotional coaching on the sub-factors of self-monitoring and self-control: self-estimation, self-determination, behavioral inhibition, and emotionality.

4.1 The Influence of Emotional Coaching on 5 Year-Old Children's Self-Regulation

4.1.1 Analysis of Homogeneity Between the Experimental Group and the Comparison Group A

As shown in Table 3, the independent sample t test, implemented to check the age difference between two groups, showed the result that the average age of the moon of the children's in the experimental group was 54.8 months old and that of the comparison group was 54.8 months old. It was verified that both groups had the same average age of the moon. Also, with the significant level of 0.05 between two groups, no significant difference was observed. Therefore, the children's of the experimental group and the comparison group can be considered as a homogenous group in terms of average age of the moon and gender.

The reliability of self-regulation at the construct level and the sub-factor level was verified. Reliability verification was executed by calculating the value of Cronbach's α that shows the internal consistency of questionnaire items. Table 3 suggests the

Table 3 A tool for measuring self-regulation

Category		Definition	Question number	Cronbach's α
Self-monitoring	Self-estimation	To recognize one's own cognitive process and the result	5	0.944
	Self-determination	The ability to select proper strategy depending on situations after estimating one's own cognitive process	9	0.924
Self-control	Behavioral inhibition	The ability to control the intensity, the frequency, the speed, and the strength of behavior under instructions	10	0.932
	Emotionality	Variable for predicting prosocial behavior and problem behavior	7	0.929
Self-regulation			31	0.978

definitions and the numbers of questionnaire items about the sub-factors of self-regulation, and the value of Cronbach's α. As the value of Cronbach's α on the variables of self-regulation and its sub-factors exceeds 0.70 (Nunnally 1978), which can be generally accepted in social science researches, it was confirmed that the reliability of each variable had no problem.

4.1.2 Analysis: Stage 1

In the research, a t-test was conducted to examine the difference verification of the scores in the pre-test and the post-test with a view to investigating the influence of emotional coaching on children's self-regulation. The findings regarding the difference in five-year-old children's self-regulation between the experimental and comparison groups are presented as follows. In the pre-test result, the average score of the experimental group ($M = 2.34$, $SD = 0.265$) and the comparison group ($M = 2.28$, $SD = 0.388$) didn't show a significant difference. However, in the post-test result, the experimental group had a higher average score of self-regulation ($M = 3.54$, $SD = 0.363$) than the comparison group ($M = 2.40$, $SD = 0.292$), and it revealed a statistically significant difference ($t = 9.882$, $p < 0.001$). It indicates that emotional coaching has a positive effect on children's self-regulation.

4.1.3 Analysis: Stage 2

The results of the t-test on self-monitoring and self-control, the sub-factors of self-regulation, are summarized as follows. As for self-monitoring, in the result of the pre-test, the average score gap between the experimental group ($M = 2.38$, $SD = 0.472$) and the comparison group ($M = 2.37$, $SD = 0.530$) wasn't statistically significant ($t = 0.047$, $p < 0.05$). However, in the result of the post-test, the experimental group showed a higher average score of self-monitoring ($M = 3.67$, $SD = 0.425$) than the comparison group ($M = 2.31$, $SD = 0.506$), and it revealed a statistically significant difference ($t = 8.404$, $p < 0.001$).

As for self-control, in the result of the pre-test, the average score gap between the experimental group ($M = 2.10$, $SD = 0.464$) and the comparison group ($M = 2.26$, $SD = 0.550$) wasn't statistically significant ($t = 1.244$, $p < 0.05$). However, in the result of the post-test, the average score of the experimental group ($M = 3.41$, $SD = 0.521$) was higher than the comparison group ($M = 2.29$, $SD = 0.385$), and it also showed a statistically significant difference ($t = 8.426$, $p < 0.001$).

4.1.4 Analysis: Stage 3

The t-test results for self-evaluation, self-determination, behavioral inhibition, and emotionality (sub-factors of self-monitoring and self-control) are summarized and presented as follows. In the pre-test on the four sub-factors, the average scores of the experimental group and the comparison group were nearly similar level, and a statistically significant difference wasn't observed. However, the result of the post-test indicated that the average scores of the four sub-factors revealed statistically significant differences between the experimental group and the comparison group.

As for self-evaluation, the average score of the experimental group was 3.75 points while that of the comparison group was 2.76 points, and therefore, the experimental group showed a higher level of self-evaluation in self-regulation, which had a statistically significant difference ($t = 6.740$, $p < 0.001$). In the case of self-determination, the average score of the experimental group was 3.60 points, comparing to 2.46 points of the comparison group, which suggests that the experimental group showed a higher level of self-determination in self-regulation, and the result suggests a statistically significant difference ($t = 4.359$, $p < 0.001$). As for behavioral inhibition, considering the experimental group with 3.25 points and the comparison group with 2.49, the experimental group showed a higher level of behavioral inhibition in self-regulation, which had a statistically significant difference ($t = 4.046$, $p < 0.01$). In the case of emotionality, the average score of the experimental group was 3.57 points while that of the comparison group was 2.29 points, which demonstrates the experimental group had a higher level of self-regulation of emotionality, and the result suggests a statistically significant difference ($t = 5.990$, $p < 0.001$).

Table 4 the results of pre-rest and post-test for the experimental group and the comparison group by sub-factor of self-regulation.

Table 4 Results of verification of differences by sub-factors of self-regulation ($N = 30$)

Category				*N*	*M*	*SD*	*t*	*p*
Self-monitoring	Self-estimation	Pre-test	Experimental group	30	2.55	0.588	0.159	0.876
			Comparison group	30	2.52	0.641		
		Post-test	Experimental group	30	3.75	0.670	6.740***	0.000
			Comparison group	30	2.76	0.638		
	Self-determination	Pre-test	Experimental group	30	2.31	0.679	0.143	0.943
			Comparison group	30	2.23	0.698		
		Post-test	Experimental group	30	3.60	0.778	4.359**	0.001
			Comparison group	30	2.46	0.614		
Self-control	Behavioral inhibition	Pre-test	Experimental group	30	2.41	0.684	0.132	0.740
			Comparison group	30	2.35	0.781		
		Post-test	Experimental group	30	3.25	0.638	4.046**	0.001
			Comparison group	30	2.49	0.599		
	Emotionality	Pre-test	Experimental group	30	2.20	0.695	0.103	0.589
			Comparison group	30	2.18	0.660		
		Post-test	Experimental group	30	3.57	0.808	5.990***	0.000
			Comparison group	30	2.29	0.670		

$^{**}p < 0.01$, $^{***}p < 0.001$

5 Conclusion and Suggestions

This study aims to verify the effect of emotional coaching on the development of the self-regulation of five-year-old children's. The result of the study intends to provide a basis that can clarify the understanding of emotional coaching strategy and help the development of emotional coaching programs that can be applied to children in the early childhood education field.

The subjects derived from the research questions and the conclusion are as the following: First, as the result of applying emotional coaching to five-year-olds, positive effects on the development of children's self-regulation were observed. And the result of comparing the scores of the pre-test and the post-test to investigate the influence of emotional coaching on the self-regulation of five-year-olds showed that the post-test scores of the experimental group were significantly high.

This study provided activities, through interactions and emotional coaching approach, like "oh, I see", or "I'm with you", designed to encourage children's to experience their emotions being accepted and display them when they face negative emotions, such as 'sadness', 'anger', and 'depression'. As these activities developed, it was observed that the activities contributed to the improvement of self-regulation by providing the children's with opportunities to express their negative emotions within the acceptable range, without prejudices on emotions. This lies in the same context as the study by Yeonju (2013), which states emotional coaching activities have a positive influence on the improvement of children's self-regulation and help children's control their own emotions, understand others' feelings and adjust their behavior to others' thoughts and opinions.

Secondly, this study confirms that emotional coaching has a positive influence on self-monitoring and self-control, the sub-factors of self-regulation. Especially, it exerted a positive impact on self-determination and self-estimation, the sub-factors of self-monitoring. This result is in line with the theory by Vygotsky (1978), which suggests that children enhance their self-regulation by controlling their feelings and behavior with their external language. Additionally, emotional coaching method applied to children's showed a positive effect on behavioral inhibition and emotionality, the sub-factors of self-control. This shared the same result as the study by Okseung (2006), which has the conclusion that children develop their self-regulation through experiences of planning, adjusting and controlling their behavior and thoughts at play, and it is partly in line with the research result by Hyeok (2002), which concludes that the activities of personality education projects advance behavioral regulation ability of children.

The result of this study confirms the importance of interactions between teachers and children. It shows that the impact of interactions between teachers and children's on the self-regulation ability of five-year-olds depends on the methods and content used during activities with the same subjects. That is, empathizing with children's emotions through emotional coaching methods and limiting their negative behavior resulted in the improvement of self-regulation. Eventually, in order to encourage children to regulate their own behavior by controlling their own emotions, strong support and interactions with adults are required (Okseung 2006).

The result of this study demonstrates that as the children's with an experience of emotional coaching watch the teachers and the peers listening to what they say and sharing how they feel and, furthermore, feel themselves being respected, they can get beyond self-centered thoughts and advance their self-regulation ability to adjust and control their own emotions as a member of the society. Besides, it also implies that applying emotional coaching, by improving children's self-regulation, can prove its educational value in the education field. And emotional coaching is anticipated to play a role in enhancing five-year-olds' self-regulation as well.

Based on limits of this study, suggestions on the follow-up study are as the following: First, although this study verified the effect of emotional coaching on self-regulation by classifying self-regulation into its sub-factors, differentiated from the previous study, the limited number of 30 samples may have restricted drawing stable outcome. Therefore, in the follow-up study, more samples are required so that it can draw more statistically stable outcome. Supposing that longitudinal research design is applied in the follow-up study, it is anticipated that the causality between two variables can be more clearly identified. Finally, as the effectiveness verification tool of this study relies on observations by teachers, it may have the possibility of errors from common method bias if a teacher's subjective assessment is involved in it. The follow-up study is required to reduce the possibility of errors from common method bias, for instance, by using multilateral evaluation methods or changing evaluators from teachers to parents.

References

Eunjoo L (2010) Analysis of recent research trend on the formation of peer relationship. Sourcebook of the symposium in Autumn. Korean Assoc Child Stud 13(1):33–48

Eunju J (2005) Factor research affecting behavioral and emotional self-regulation of children's. Doctoral thesis, Graduate School of Seoul Women's University

Eunyoung N (2007) Emotional coaching on art of love for my child. The Korea Economic Daily, Seoul

Gettman HJ, Gelfand MJ (2007) When the customer shouldn't be king: antecedents and consequences of sexual harassment by clients and customers. J Appl Psychol 92(3):757

Gottman JM, Seongae C, Byeok C (2011) Emotional coaching for my child. The Korea Economic Daily, Seoul

Grand A (2003) When the show must go on: surface acting and deep acting as determinants of emotional exhaustion and peer-rated service delivery. Acad Manag J 46(1):86–96

Hyeok H (2002) The influence of personality education project with computers on children's self-regulation. Doctoral thesis, The Graduate School of Donga University

Hyesoon C (2011) Child guidance. Shinjeong, Seoul

Hyonkyong L (2012) The influence of drama activities using personality-education-related picture books on prosocial behavior and emotional intelligence of children's. Master's thesis, The Graduate School of Ehwa Women's University

Jeongran L (2003) Structural analysis of component factors of self-regulation of children's and related variables. Doctoral thesis, Graduate School of Duksung Women's University

Jihye K (2015) The effect of emotional coaching on friendship of elementary school students. Doctoral thesis, Graduate School of Education of Geyongsang University

Jihyeon J (2010) The influence of self-directed history education activities on children's self-regulation. Doctoral thesis, Graduate School of Duksung Women's University

Jihyeon O (2016) The influence of mother's experiences of acceptance in childhood, social support, and reactions of emotional coaching on the way of children's emotional regulation. Acad J 37(4)

Mihwa J (2013) The influence of personality education associated with the home on children's self-regulation and parents' recognition about personality education. Master's thesis, The Graduate School of Education of Yonsei University

Misook J (2016) The effect of programs with emotional coaching on the improvement of children's self-regulation. Master's thesis, Graduate School of Mokpo University

Nunnally (1978) Psychometric Theory, 2nd edition, New York, McGraw-Hill

Okseung Y (2006) The development of measurement scales of self-regulation in children's at the age of 3 to 6. Acad J

Seongae C, Byeok C (2012) Emotional coaching for teenagers by Choi Seongae and Cho Byeok. Hainaim, Seoul

Seongae M (2012) The effect of emotional coaching on relationship formation and communication of children's with emotional and behavioral problems. Master's thesis, Graduate School of Education of Soongsil University

Seongeui P, Mihyang D (2019) The development and the effect of parental coaching program for mothers with children of early childhood. Korean Coach Assoc 16(1):45–67

Seongeun L (2013) The influence of personality education activities using picture books on self-regulation and prosocial behavior of five-year-olds. Master's thesis, Graduate School of Education of Sungshin Women's University

Vygotsky LS (1978) Consciousness and proficiency. Salimteo

Yeonju P (2013) The influence of emotional coaching on self-regulation, self-esteem, and interpersonal relationship in elementary school students. Master's thesis, Graduate School of Education of Geyongsang University

Yooni K (2013) Research on lesson plans for emoticon design with emotional coaching in visual art education. Doctoral thesis, Graduate School of Education of Kookmin University

Younghee S (2014) The influence of educational programs in early childhood using emotional coaching on children's sociality. Doctoral thesis, Graduate School of Education of Kyeimyung University

Yujin Y (2010) The impact of conversations using picture books on children's self-regulation. Master's thesis, Graduate School of Duksung Women's University

retical Background

erpersonal Competence

nal competence is the process of recognizing and responding to others, rucial effect on relationship formation, maintenance, and social interaction ve interpersonal relationships in the workplace have an effect on enhancing s' intrinsic motivation and improving overall organizational performance

ersonal competence is divided into communication ability, group inter-relationships, and daily life interpersonal relationships [8]. Communica-ty refers to relationship formation through verbal and non-verbal inter-roup interpersonal relationships refer to harmonious goal achievement collaboration and teamwork, and daily life interpersonal relationships refer thening social bonds through daily interactions. These abilities are consid-ntial psychological and social competencies that have an effect on individual gical stability and organizational productivity, serving as vital resources for al social adaptation and organizational role performance and effectiveness ment [8].

elf-Efficacy

icacy represents an individual's conviction regarding their capability to effec-xecute behaviors necessary in particular circumstances. It can be concep-l as one's self-perceived ability to systematically arrange and implement d responsibilities [9]. Research indicates that individuals possessing elevated icacy levels demonstrate increased motivation and psychological well-being, uently generating favorable outcomes in organizational contexts [10]. Self-y functions as a crucial psychological asset that enables adaptable manage-f stressful circumstances and strengthens problem-solving capabilities, poten-nhancing individual productivity and sustainability across diverse workplace s [11].

Psychological Well-Being

ological well-being is viewed as a positive indicator of mental health that ents the degree of happiness and satisfaction experienced in daily life and ersonal relationships [12]. Ryff structured psychological well-being into six lements: self-realization, personal growth, purpose in life, autonomy, environ-al mastery, and positive relationships [13]. This includes the process of positively

A Study on the Effect of Inte Competence on Turnover Int Focusing on Self-Efficacy, Ps Well-Being, and Organizatioı Commitment

Sei-Yon Hong and Jong-Woo Park

1 Introduction

The concept of lifelong employment is gradually wea turnover across organizational boundaries is becomiı employee turnover has a close relationship with psych such as organizational culture, interpersonal relationsł

Interpersonal competence, as a social skill that effe with others, has a significant effect on organizational ad Positive interpersonal relationships within organization self-efficacy [4]. Organizational commitment is a key intention, and higher levels of commitment tend to have nizational performance [5]. As the significance of psycho influencing turnover intention becomes more prominent, integrated structural relationships among these variables

This study aims to verify the structural relationsh competence and turnover intention through psychologic efficacy, psychological well-being, and organizational c that establishing psychology-based management strategi relationships within organizations can have an effect o efficiency.

S.-Y. Hong
Graduate School of Project Management, Soongsil University, Seoul,
e-mail: soyouni1022@naver.com

J.-W. Park (✉)
Department of Business Administration, Soongsil University, Seoul, R
e-mail: Jongpark7@ssu.ac.kr

R. Lee et al. (eds.), *IT Service & Business Innovation*, Studies in Comp
Intelligence 1252, https://doi.org/10.1007/978-3-032-14397-6_8

2 Theo

2.1 Int

Interpers having a [6]. Posit employee [7].

Interp personal tion abil action, through to streng ered ess psychol individu enhance

2.2 S

Self-ef tively tualize assign self-ef subseq efficac ment c tially settin

2.3

Psyc repre inter core men

evaluating one's life, continuous growth, and maintaining identity and life direction. Ultimately, psychological well-being can be viewed as a core psychosocial resource that has an effect on individual mental health and overall quality of life [13].

2.4 Organizational Commitment

Organizational commitment refers to the psychological state where employees agree with organizational goals and values and desire to contribute to the organization's maintenance and development [14]. Meyer and Allen categorized organizational commitment into three dimensions based on how organizational members form and maintain psychological bonds with the organization. First, affective commitment refers to attachment and identification with the organization, second, continuance commitment refers to cost–benefit calculations of staying versus leaving, and third, normative commitment refers to moral obligation to the organization [15]. Organizational commitment has an effect on employees' job attitudes and behaviors and is used as a concept to predict and enhance organizational effectiveness [16].

2.5 Turnover Intention

Turnover intention is the psychological tendency of employees to voluntarily leave their workplace, combining the behavioral element of 'turnover' and the motivational element of 'intention' [17]. This is generally considered in relation to current job environment and includes activities such as searching for other jobs or contemplating turnover [18]. Such turnover intention is an important human resource management variable, and high turnover intention requires strategic management as it has an effect on increasing turnover rates, decreasing organizational commitment and performance, and increasing replacement workforce training costs [19].

3 Research Method

3.1 Research Method and Hypothesis Setting

This study aimed to examine the structural relationships between interpersonal competence and turnover intention through self-efficacy, psychological well-being, and organizational commitment among corporate employees. The research model is shown in Fig. 1.

Previous studies suggest that interpersonal relationships influence self-efficacy and psychological well-being in organizational settings. Research has demonstrated

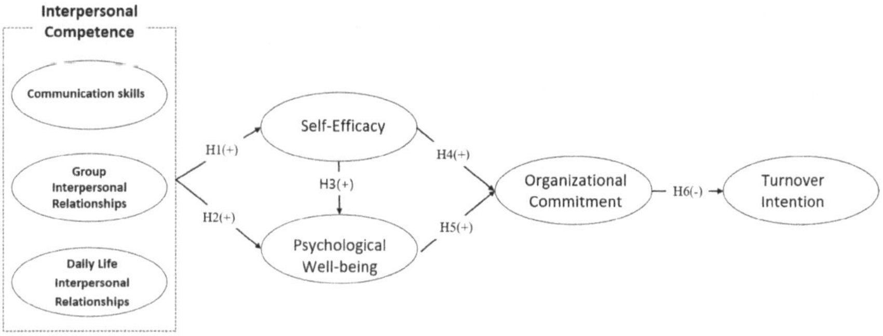

Fig. 1 Research model

that interpersonal competence components—including communication ability, group dynamics, and daily social interactions—contribute significantly to enhanced self-efficacy levels [20] and demonstrate substantial positive impacts on psychological well-being [8, 21]. Furthermore, research indicates that self-efficacy serves as a crucial determinant positively affecting both individual psychological well-being and organizational commitment [22]. Studies have also established that psychological well-being significantly influences turnover intention through its effect on organizational commitment [23]. Drawing from this theoretical foundation, we propose the following hypotheses:

H1: Interpersonal competence will have a positive (+) effect on self-efficacy.

H1-1: Communication skill will have a positive (+) effect on self-efficacy.
H1-2: Group interpersonal relationships will have a positive (+) effect on self-efficacy.
H1-3: Daily life interpersonal relationships will have a positive (+) effect on self-efficacy.

H2: Interpersonal competence will have a positive (+) effect on psychological well-being.

H2-1: Communication skill will have a positive (+) effect on psychological well-being.
H2-2: Group interpersonal relationships will have a positive (+) effect on psychological well-being.
H2-3: Daily life interpersonal relationships will have a positive (+) effect on psychological well-being.

H3: Self-efficacy will have a positive (+) effect on psychological well-being.
H4: Self-efficacy will have a positive (+) effect on organizational commitment.
H5: Psychological well-being will have a positive (+) effect on organizational commitment.
H6: Organizational commitment will have a negative (-) effect on turnover intention.

3.2 *Data Collection and Analysis Method*

This study collected data from 369 employees through a survey conducted from August 11 to August 26, 2023, targeting 303 working professionals. Statistical analysis was performed utilizing SPSS 24.0 and Smart PLS 4.0. To ensure reliability and validity, measurement model fitness was verified through exploratory factor analysis. Subsequently, structural model verification and hypothesis testing were con-ducted to verify the research model fitness.

3.3 *Measurement Tools*

The questionnaire items were organized based on previous research and literature review to test the research hypotheses. The operational definitions and measurement instruments for the variables are presented in Table 1.

4 Research Results

4.1 *Measurement Model Analysis*

Validation of the measurement model was conducted using SPSS 24.0 and Smart PLS 4.0. In this study, a reflective measurement model was built, and the validation criteria included internal consistency reliability, convergent validity, and discriminant validity.

4.1.1 Internal Consistency Reliability

Internal consistency reliability is defined as the degree of consistency among the measurement items. The results revealed that all values for Cronbach's alpha were above 0.6, all values for rho_A exceeded 0.7, and all composite reliability (CR) values were greater than the 0.6 threshold. Therefore, it can be concluded that all latent variables exhibited sufficient internal consistency reliability. The detailed results are presented in Table 2.

4.1.2 Convergent Validity

Convergent validity indicates the extent to which a latent variable represents the same concept, and is assessed using outer loadings, indicator reliability, and average variance extracted (AVE). In this study, the outer loadings ranged from 0.704 to

Table 1 Operational definitions of variables

Variable		Operational definition	References
Interpersonal competence	Communication skills	The ability to understand and execute effective communication, achieve personal goals through successful communication, and demonstrate appropriate verbal and non-verbal behavior in given situations	Schlien and Guerney [24] Choi [8] Lee and Park [20]
	Group interpersonal relationships	The ability to cooperate and interact harmoniously with other members in one's group, and maintain relationships through appropriate and effective communication within the group	
	Daily life interpersonal relationships	The ability to establish and maintain relationships with others, and effectively regulate verbal and non-verbal behavior with others	
Self-efficacy		One's belief in their ability to successfully and efficiently perform tasks in the workplace	Chen et al. [10]
Psychological well-being		Defined as an individual's subjective life satisfaction and positive emotions	Ryff [13]
Organizational commitment		The degree to which organizational members have intentions to leave their organization	Meyer and Allen [15] Mobley [25]
Turnover intention		The degree to which organizational members have intentions to leave their organization	Wayne et al. [17]

0.909, meeting the recommended threshold of 0.7 or higher. Indicator reliability values ranged from 0.501 to 0.826, all exceeding the minimum requirement of 0.5. In addition, all AVE values were above the threshold of 0.5, demonstrating that convergent validity was established. The detailed results are presented in Table 3.

4.1.3 Discriminant Validity

Assessment of discriminant validity entails evaluating the degree of distinction between latent variables through various methodological approaches. The established evaluation criteria encompass cross loadings examination, heterotrait-monotrait ratio (HTMT) analysis, and application of the Fornell-Larcker criterion. Within the framework of this investigation, measurement item outer loadings demonstrated superior values compared to their corresponding cross loadings, thereby validating appropriate item loading onto respective latent variables. The HTMT analysis revealed that all values between latent variables remained consistently below the critical

Table 2 Internal consistency reliability results

Latent variable		Measurement variable	Internal consistency reliability		
			Cronbach's a	rho_A	CR
			≥0.60	≥0.70	≥0.60
Interpersonal competence	Communication skills	CS1	0.828	0.864	0.896
		CS2			
		CS3			
	Group interpersonal relationships	GR2	0.830	0.840	0.887
		GR3			
		GR4			
		GR6			
	Daily life interpersonal relationships	DR2	0.805	0.837	0.884
		DR3			
		DR4			
Self-efficacy		SE1	0.941	0.941	0.951
		SE2			
		SE3			
		SE4			
		SE5			
		SE6			
		SE7			
		SE8			
Psychological well-being		PW7	0.886	0.887	0.913
		PW8			
		PW9			
		PW16			
		PW17			
		PW18			
Organizational commitment		OC13	0.865	0.911	0.907
		OC14			
		OC15			
		OC16			
		OC17			
		OC18			
Turnover intention		TI2	0.806	0.906	0.878
		TI3			
		TI4			

Note CS: Communication skills, GR: Group Interpersonal Relationships, DR: Daily Life Interpersonal Relationships, SE: Self-Efficacy, PW: Psychological Well-being, OC: Organizational, Commitment, TI: Turnover Intention

Table 3 Convergent validity results

Latent variable		Measurement variable	Convergent validity			Discriminant validity
			Outer loading rerevance	Reliability	AVE	HTMT
			≥0.70	≥0.50	≥0.50	Confidence interval does not include 1
Interpersonal competence	Communication skills	CS1	0.909	0.826	0.743	Not included
		CS2	0.897	0.804		
		CS3	0.774	0.599		
	Group interpersonal relationships	GR2	0.833	0.693	0.662	Not included
		GR3	0.848	0.719		
		GR4	0.833	0.693		
		GR6	0.735	0.540		
	Daily life interpersonal relationships	DR2	0.897	0.804	0.718	Not included
		DR3	0.876	0.767		
		DR4	0.763	0.582		
Self-efficacy		SE1	0.755	0.570	0.708	Not included
		SE2	0.838	0.702		
		SE3	0.825	0.680		
		SE4	0.842	0.708		
		SE5	0.894	0.799		
		SE6	0.880	0.774		
		SE7	0.857	0.734		
		SE8	0.835	0.697		
Psychological well-being		PW7	0.842	0.708	0.638	Not included
		PW8	0.733	0.537		
		PW9	0.788	0.620		
		PW16	0.800	0.640		
		PW17	0.826	0.682		
		PW18	0.800	0.640		
Organizational commitment		OC13	0.846	0.716	0.620	Not included
		OC14	0.803	0.645		
		OC15	0.704	0.501		
		OC16	0.858	0.736		
		OC17	0.740	0.547		
		OC18	0.760	0.578		
Turnover intention		TI2	0.742	0.550	0.707	Not included

(continued)

Table 3 (continued)

Latent variable	Measurement variable	Convergent validity			Discriminant validity
		Outer loading rerevance	Reliability	AVE	HTMT
		≥0.70	≥0.50	≥0.50	Confidence interval does not include 1
	TI3	0.888	0.788		
	TI4	0.884	0.781		

Note CS: Communication skills, GR: Group Interpersonal Relationships, DR: Daily Life Interpersonal Relationships, SE: Self-Efficacy, PW: Psychological Well-being, OC: Organizational, Commitment, TI: Turnover Intention, HTMT: Heterotrait-Monotrait Ratio

threshold of 0.85, with confidence intervals excluding 1, thus satisfying discriminant validity requirements. Furthermore, the Fornell-Larcker criterion analysis indicated that the correlation coefficient between self-efficacy and daily life interpersonal relationships (0.613) was demonstrably lower than the minimum square root of AVE (0.787), providing additional confirmation of discriminant validity. A comprehensive presentation of these analytical findings is detailed in Table 4.

4.2 Hypothesis Testing

Based on the validation of the reflective measurement model, the reliability and validity between the measurement items and latent variables were confirmed; thus, a structural model analysis was conducted. The structural model was evaluated using bootstrapping to determine the significance of the path coefficients. Generally, a t-value of ±1.96 or higher, or a p-value less than 0.05, was considered statistically significant.

This study analyzed the structural relationships in which interpersonal competence affects turnover intention, mediated by self-efficacy, psychological well-being, and organizational commitment. Of the ten hypotheses tested, eight were supported, except for H1-1 and H2-1. The detailed results are presented in Table 5.

5 Conclusion

This study examined the effects of interpersonal competence sub-variables (communication, group relations, and daily relations) on turnover intention among organizational members. The structural relationships between interpersonal competence and

Table 4 Evaluation result of discriminant validity according to the Fornell-Larcker criterion

	PW	CS	TI	DR	SE	OC	GR
PW	*0.799*						
CS	0.406	*0.862*					
TI	–0084	–0163	*0.841*				
DR	0.580	0.555	−0.145	*0.847*			
SE	0.062	0.404	−0.091	0.613	*0.842*		
OC	0.290	0.212	−0.350	0.313	0.283	*0.787*	
GR	0.481	0.589	−0.138	0.581	0.480	0.243	*0.814*

* The diagonally shaded cells are the square root of AVE, and the rest are the values of the correlation co efficient

Note CS: Communication skills, GR: Group Interpersonal Relationships, DR: Daily Life Interpersonal Relationships, SE: Self-Efficacy PW: Psychological Well-being, OC: Organizational Commitment, TI: Turnover Intention

Table 5 Path coefficient analysis results of the structural model (* overall sample (N = 369))

	Hypothesized paths			PC	SD	T	P	Result
H1	Interpersonal competence → self-efficacy							
H1-1	Communication skills	→	Self-efficacy	0.024	0.069	0.35	0.726	Reject
H1-2	Group interpersonal relationships	→	Self-efficacy	0.177	0.076	2.329	0.02*	Accept
H1-3	Daily life interpersonal relationships	→		0.497	0.064	7.718	0.000***	
H2	Interpersonal competence → psychological well-being							
H2-1	Communication skills	→	Psychological well-being	0.036	0.067	0.539	0.59	Reject
H2-2	Group interpersonal relationships	→	Psychological well-being	0.135	0.069	1.973	0.049*	Accept
H2-3	Daily life interpersonal Relationships	→		0.257	0.076	3.38	0.001**	
H3	Self-efficacy	→	Psychological well-being	0.364	0.069	5.297	0.000***	Accept
H4	Self-efficacy	→	Organizational commitment	0.17	0.073	2.342	0.019**	Accept
H5	Psychological well-being	→		0.188	0.069	2.717	0.007**	Accept
H6	Organizational commitment	→	Turnover intention	-0.35	0.05	7.708	0.000***	Accept

** p0.001, ** p0.01, * p0.05 >>> *** p0.001
Note PC: Path Coefficient, SD: Standard Deviation, T: T Value, P: P Value

turnover intention were verified by establishing self-efficacy, psychological well-being, and organizational commitment as variables. The results and implications of this study are as follows.

5.1 Research Results

First, testing of interpersonal competence sub-factors revealed that both group relations and daily relations showed significant positive effects on self-efficacy and psychological well-being. However, communication skills did not demonstrate significant effects on either self-efficacy or psychological well-being. Second, self-efficacy demonstrated a positive effect on psychological well-being. Third, both self-efficacy and psychological well-being were verified to have effects that enhanced levels of organizational commitment. Fourth, organizational commitment was confirmed to significantly reduce turnover intention.

The finding that communication skills did not significantly affect self-efficacy in this study may be attributed to organizational characteristics, such as size and hierarchical structure. Previous research highlights that the effectiveness of communication is more influenced by qualitative factors and environmental aspects than by skills themselves, and that "authenticity" demonstrates a significant correlation [26]. Communication impacts self-efficacy when combined with individual internal and relational capacities [27], and the importance of a free and open communication environment has also been emphasized [28]. Therefore, it can be concluded that the effectiveness of communication within organizations is determined by a complex interplay of environmental and qualitative factors, in addition to individual skills.

The finding that Communication skills do not significantly affect Psychological Well-being is consistent with some existing studies [21, 29] which suggest that the vertical, task-oriented communication environment within organizations can restrict individuals' free communication, and that various communication styles due to generational changes can also have an impact. Furthermore, considering the influence of Group Interpersonal Relationships and shared goals on individual psychology [30], the effectiveness of organizational communication can vary depending on the situation. Therefore, the relationship between communication and Psychological Well-being should be understood comprehensively in the organizational context.

5.2 *Academic Implications*

This study established interpersonal competence as an independent variable, verified the structural path leading to turnover intention, and presented academic validity for enhancing individual internal factors and psychological intervention within organizations [31]. Furthermore, it provided a differentiated perspective from existing management variable-centered research through an integrated approach that predicts turnover intention from psychological variables (emotion, attitude) [32]. This holds significance in emphasizing the importance of creating emotional and psychological environments in organizations and establishing empirical foundations for the development of related academic fields.

5.3 *Practical Implications*

This study identified the mechanism by which interpersonal competence influences turnover intention through psychological changes in organizational members. This suggests the importance of introducing interpersonal relationship strategies and related training that consider individual characteristics for effective human resource management. In particular, qualitative elements such as 'authenticity' and creating an environment based on psychological safety are more important than communication ability itself [26], which contributes to improving organizational relationships and work efficiency. Improving the quality of interpersonal relationships and communication leads to practical outcomes, such as stronger organizational commitment and a lower turnover rate.

5.4 *Limitations and Suggestions for Future Research*

Future research should attempt to increase generalizability by expanding to diverse subjects (gender, age, job groups) and attempt qualitative research to complement self-report bias. Additionally, path analysis of interpersonal competence subcomponents and identification of additional variables are needed to expand the research model, and longitudinal research design is required to understand changes over time.

References

1. Cooper-Hakim A, Viswesvaran C (2005) The construct of work commitment: testing an integrative framework. Psychol Bull 131(2):241–259
2. Sung SH, Min HJ (2020) Study on the concept map of a good job perceived by employees who changed jobs. Korean J Employ Career 10(3):49–70
3. Spitzberg BH, Cupach WR (2012) Handbook of interpersonal competence research. Springer Science & Business Media
4. Hong SY, Cho DH (2023) A study on the impact of adult attachment on organizational commitment among service workers. J Korea Acad-Ind Coop Soc 24(4):465–476
5. Petrou P, Demerouti E, Peeters MCW, Schaufeli WB, Hetland J (2012) Crafting a job on a daily basis: contextual correlates and the link to work engagement. J Organ Behav 33(8):1120–1141
6. Heider F (2013) The psychology of interpersonal relations. Psychology Press
7. Mossholder KW, Bedeian AG, Armenakis AA (1982) Group process–work outcome relationships: a note on the moderating impact of self-esteem. Acad Manag J 25(3):575–585
8. Choi YM (2019) The effect of leisure activity satisfaction on psychological well-being in elderly users of social welfare facilities: focusing on the mediating effect of interpersonal relationships and the moderating effect of ego resilience. Doctoral dissertation, Sungkyul University
9. Bandura A (1977) Self-efficacy: toward a unifying theory of behavior change. Psychol Rev 84(2):191–215
10. Chen G, Gully SM, Eden D (2001) Validation of a new general self-efficacy scale. Organ Res Methods 4(1):62–83
11. Luszczynska A, Scholz U, Schwarzer R (2005) The general self-efficacy scale: multicultural validation studies. J Psychol 139(5):439–457
12. Keyes CL, Shmotkin D, Ryff CD (2002) Optimizing well-being: the empirical encounter of two traditions. J Pers Soc Psychol 82(6):1007–1022
13. Ryff CD (1995) Psychological well-being in adult life. Curr Dir Psychol Sci 4(4):99–104
14. Simon HA (1959) Theories of decision-making in economics and behavioral science. Am Econ Rev 49(3):253–283
15. Meyer JP, Allen NJ (1997) Commitment in the workplace: theory, research, and application. Sage Publications

16. Allen NJ, Meyer JP (1990) The measurement and antecedents of affective, continuance and normative commitment to the organization. J Occup Psychol 63(1):1–18
17. Wayne SJ, Shore LM, Liden RC (1997) Perceived organizational support and leader-member exchange: a social exchange perspective. Acad Manag J 40(1):82–111
18. Mowday RT, Steers RM, Porter LW (1979) The measurement of organizational commitment. J Vocat Behav 14(2):224–247
19. Hom PW, Caranikas-Walker F, Prussia GE, Griffeth RW (1992) A meta-analytical structural equations analysis of a model of employee turnover. J Appl Psychol 77(6):890–909
20. Lee CI, Park KN (2020) The effect of participation in exercise programs on successful aging mediated by interpersonal ability in the elderly. Glob Creat Lead 10(1):303–325
21. Kim DJ, Choi CJ (2020) The influence of leisure activity satisfaction on psychological well-being mediated by self-efficacy and interpersonal relationship ability in the elderly. Korean J Hum Welf Pract Res 24:17–115
22. Cayupe JC, Bernedo-Moreira DH, Morales-García WC, Alcaraz FL, Peña KBC, Saintila J, Flores-Paredes A (2023) Self-efficacy, organizational commitment, workload as predictors of life satisfaction in elementary school teachers: the mediating role of job satisfaction. Front Psychol 14, Article 1066321
23. Hong SY (2023) A study on the effect of interpersonal relationship ability on turnover intention. Doctoral dissertation, Soongsil University
24. Schlein S, Guerney Jr BG, Stover L (1971) The interpersonal relationship scale. Unpublished doctoral dissertation, University of Pennsylvania
25. Mobley WH (1977) Intermediate linkages in the relationship between job satisfaction and employee turnover. J Appl Psychol 62(2):237–240
26. Kim SH (2006) Relationship between facilitative communication patterns, communication competence, and self-efficacy among psychiatric nurses. Master's thesis, Korea University
27. Hsieh HF, Chen KM, Lin MH, Wang YC, Huang HT (2014) Social welfare utilization and needs of older adults in Taiwan: survey research. Nurs Outlook 62(6):459–468
28. Clark RD, Shields G (1997) Family communication and delinquency. Adolescence 32(125):81–93
29. Lee BS (2023) The effect of family health perceived by middle-aged adults on psychological well-being. Doctoral dissertation, Calvin University
30. Asch SE (1955) Opinions and social pressure. Sci Am 193(5):31–35
31. Ko HJ (2023) Analysis of research trends on interpersonal competence. J Couns Educ 6(1):1–16
32. Gregory DM, Way CY, Le Fort S, Barrett BJ, Parfrey PS (2007) Predictors of registered nurses' organizational commitment and intent to stay. Health Care Manag Rev 32(2):119–127

The Effect of a Leader's Coaching Behavior on Employees' Quiet Quitting Through Work Engagement and Job Satisfaction

Seong-Sil Jeong and Jong-Woo Park

1 Introduction

A Gallup survey of 15,091 American adult workers found that a significant number of respondents were emotionally disconnected from their organizations due to low work engagement [1]. While maintaining formal employment relationships, these individuals showed markedly decreased enthusiasm and participation in their work. This quiet quitting phenomenon can be interpreted as a structural issue affecting the entire organization rather than just an individual problem. It refers to employees not officially leaving their jobs but performing only minimal duties commensurate with their compensation, without demonstrating the effort or commitment expected by the organization [2]. The causes of quiet quitting are closely related to factors such as job burnout, leadership quality, organizational culture, and perceptions of fairness. Among these, the lack of attention and support from leaders is particularly noted as a key factor triggering quiet quitting. Harter suggested a link wherein employees' feelings of being uncared for and disrespected by their leadership correlate with a decline in emotional attachment to the organization, consequently leading to quiet quitting. Suggested that employees who do not feel cared for and respected by their leaders tend to show decreased emotional commitment to the organization, leading to quiet quitting [1]. Leaders' indifference accelerates employees' psychological withdrawal, ultimately resulting in decreased organizational commitment and productivity [3]. Against this background, this study aims to systematically explore factors related to quiet quitting, particularly analyzing the relationships between a leader's coaching behavior, work engagement, job satisfaction, and quiet quitting. Through this, we

S.-S. Jeong
Graduate School of Project Management, Soongsil University, Seoul, Republic of Korea
e-mail: yhhhaaa@hanmail.net

J.-W. Park (✉)
Department of Business Administration, Soongsil University, Seoul, Republic of Korea
e-mail: Jongpark7@ssu.ac.kr

R. Lee et al. (eds.), *IT Service & Business Innovation*, Studies in Computational Intelligence 1252, https://doi.org/10.1007/978-3-032-14397-6_9

will deepen our understanding of quiet quitting and derive practical implications for effective management strategies.

2 Theoretical Background

2.1 Coaching Behavior

Coaching is defined as a process that promotes growth by utilizing individual resources and positively influences member performance through leader feedback [4]. It focuses on helping individuals develop their learning abilities and realize their potential beyond simple instructions. Additionally, coaching is emphasized as an interactive dialogue where organizational members and leaders share common goals and show consistent behavior, highlighting the relational aspect as the interaction between managers and subordinates for performance improvement [5]. This perspective suggests that coaching is not merely knowledge transfer, but a method for members to develop their learning capacity, enhance motivation and abilities, and maximize potential [6]. In organizational environments, a manager's coaching behavior is gaining attention as a meaningful alternative for member participation and performance creation within horizontal structures [7] and refers to the process of supporting and motivating members to develop hidden capabilities and maximize their abilities [8]. It is specific supporting behavior provided by managers as coaches [9] and is considered both a routine work process that provides opportunities for improving member performance and job skills, and a unique leadership style [10].

2.2 Quiet Quitting

Quiet quitting refers to a state where individuals physically remain in the organization but perform only minimal job requirements due to decreased psychological and emotional engagement. This presents a critical perspective as an extension of negative work attitudes such as work avoidance, laziness, and negligence. It points out that when some members exhibiting quiet quitting behavior perform only minimal assigned tasks, additional work burdens may be transferred to other members [11]. The characteristics of members showing quiet quitting behavior carry meaning beyond reduced workload in that they maintain psychological distance from assigned work while performing only the minimum requirements demanded by the organization [12]. Another perspective on quiet quitting includes situations where employers unintentionally cause members to psychologically distance themselves from the organization by failing to properly identify necessary elements for organizational members and provide appropriate support, such as coaching and career development [13].

2.3 Work Engagement

Work engagement is an important construct explaining an individual's cognitive and emotional responses to their job [14], including positive emotions and evaluations about one's job, and is used as a key variable for job performance [15]. Work engagement is characterized as a psychological state concerning work that is positive and fulfilling psychological state concerning work, comprising three sub-components: vigor, dedication, and absorption [14]. Members with high work engagement tend to form their identity through their work, thus voluntarily investing considerable energy and effort into their tasks [16]. Members with high levels of work engagement tend to set challenging goals and actively work to achieve them, beyond simply fulfilling given tasks [17].

2.4 Job Satisfaction

Job satisfaction represents an individual's psychological state and reflects personal perspectives on oneself, the organization, and one's work [18]. This job satisfaction is recognized as one of the important variables that is not simply limited to internal emotional responses but is also significantly influenced by work environment characteristics [19]. It suggests that job satisfaction can be a reference point for forming positive perspectives on work, encompassing joy and positive emotional states about one's job [20]. Positive relationships among organizational members contribute to improving job satisfaction, and particularly, mutual trust with colleagues and superiors mediates the relationship between organizational member relationships and job satisfaction [21].

3 Research Method

3.1 Research Method and Hypothesis Setting

The purpose of this empirical investigation was to identify the role of work engagement and job satisfaction as mediators in the relationship between a leader's coaching behavior and employee quiet quitting. Additionally, the study aimed to clarify the comprehensive structural relationships existing between coaching behavior, quiet quitting, work engagement, and job satisfaction [1, 3]. As one of the various factors that cause Quiet quitting behavior, the lack of interest and support shown by the leader to the members of the organization was pointed out as a key factor triggering Quiet quitting [1, 2]. In previous studies, the leader's coaching behavior refers to the process of supporting and eastern engagement to develop the members' competencies and maximize their abilities [8] and the leader's coaching behavior suggests

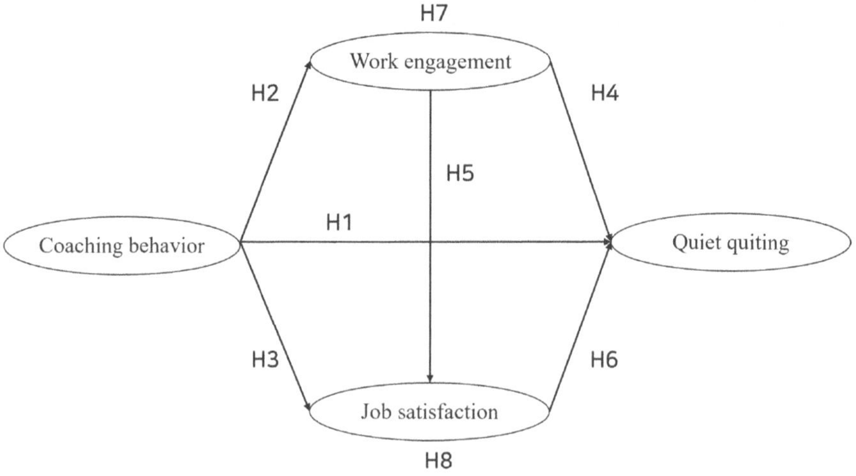

Fig. 1 A model showing the relationships among coaching behavior, quiet quitting, work engagement, and job satisfaction

that members are passionate, immersed, and committed to their jobs by improving their intrinsic motives such as job meaning and why they should perform their duties [22]. This study aims to clarify the role of relief agents in coaching behavior, Quiet quitting, work engagement, and job satisfaction within this relationship. Based on the preceding discussion, the research model for this study is depicted in (Fig. 1).

The background for establishing the research hypothesis is as follows. Based on the literature review of previous studies, this study proposes a hypothesis to explore the relationship between the leader's coaching behavior, Quiet asking, work engagement, and job satisfaction. The effect of the leader's coaching behavior on Quiet asking is examined [8, 9]. The effect of the leader's coaching behavior on job enthusiasm and job satisfaction is identified and examined [22]. The role of work engagement and job satisfaction in the relationship between the leader's coaching behavior and Quiet asking is examined [17, 21]. The specific hypothesis is as follows.

H1: A Leader's coaching behavior will have a negative (−) effect on employees' quiet quitting.

H2: The leader's coaching behavior will have a positive (+) effect on employees' work engagement.

H3: The leader's coaching behavior will have a positive (+) effect on employee job satisfaction.

H4: Work engagement will have a negative (−) effect on quiet quitting.

H5: Work engagement will have a positive (+) effect on job satisfaction.

H6: Job satisfaction will have a negative (−) effect on quiet quitting.

H7: Work engagement will mediate the relationship between a leader's coaching behavior and quiet quitting.

H8: Job satisfaction will mediate the relationship between a leader's coaching behavior and quiet quitting.

By testing these hypotheses, this study aims to deepen understanding of how a leader's coaching behavior affects an employee's quiet quitting behavior, and to identify structural relationships between coaching behavior, work engagement, and job satisfaction to find effective ways to mitigate quiet quitting tendencies and improve levels of work engagement and job satisfaction among organizational members.

3.2 Data Collection and Analysis Method

This study collected data from employees currently working in companies to empirically verify the research model presented earlier. The data was analyzed using SPSS (Statistical Package for the Social Sciences) 22.0 program to perform statistical analysis and test research hypotheses.

3.3 Measurement Tools

To test the hypotheses of the study, the operational definitions of various variables and measurement tools were organized based on prior research and literature review, Based on the preceding discussion, the research model for this study is depicted in (Table 1).

4 Research Results

4.1 Reliability and Validity Analysis

The exploratory factor analysis performed using SPSS showed that the main variables in this study each had distinct factor structures. Factors were extracted using principal component analysis, and a varimax rotation was subsequently conducted. The KMO (Kaiser–Meyer–Olkin) sampling adequacy measure was 0.959, which was excellent, and Bartlett's sphericity test result $\chi^2 = 9880.422(\text{df} = 741, p < 0.001)$ was statistically significant, confirming that the data was suitable for factor analysis.

Table 1 Operational definitions of variables

Variable	Operational definition	
Coaching behavior	Leader or manager's supportive behavior through coaching that provides psychological support, recognition, and feedback for member capacity development and performance improvement as a form of leadership	Jo Eun-Hyun and Tak Jin-Guk [11], Whitmore [12], Ellinger and Keller [13]
Quiet quitting	An attitude of psychologically distancing oneself from the organization and work while performing only the minimum required tasks within the organization, without actually leaving	Klotz and Bolino [11], Oh Hyun-Soo [12], Wigert [13]
Work engagement	A cognitive and emotional commitment state towards work, where one maintains a positive attitude and feels passionate about their job	Bakker and Schaufeli [15–17]
Job satisfaction	An employee's favorable attitude towards their job results from emotional responses, whether positive or negative, regarding their work duties	Omarov [18], Skaalvik and Skaalvik [20], Bulińska-Stangrecka and Bagieńska [21]

4.2 Correlation Analysis

The correlation analysis of the leader's coaching behavior, quiet quitting, work engagement, and job satisfaction in this study is presented in (Table 2).

5 Hypothesis Testing

The results of this study are as follows. Leader's coaching behavior was found to significantly enhance the level of work engagement ($\beta = 0.647$, $p < 0.001$). This suggests that a leader's coaching behavior supports contributes to increasing member commitment and passion. Additionally, the leader's coaching behavior was confirmed

Table 2 Correlation analysis results

Variable	Average	SD	Coaching behavior	Work engagement	Job satisfaction	Quiet quitting
Coaching behavior	4.958	1.133	1			
Work engagement	4.546	1.293	0.647[a]	1		
Job satisfaction	4.622	1.361	0.719[a]	0.779[a]	1	
Quiet quitting	3.167	1.161	−0.457[a]	−0.503[a]	−0.513[a]	1

[a] $p < 0.001$

to be a strong predictor variable that has a significant positive (+) effect on job satisfaction ($\beta = 0.719$, $p < 0.001$). Examining the relationship with the dependent variable, quiet quitting, the leader's coaching behavior showed a significant negative (−) effect in reducing quiet quitting tendencies ($\beta = -0.457$, $p < 0.001$). This means that supportive coaching by leaders directly inhibits members' quiet quitting behavior ($R^2 = 0.208$). The hypothesis testing results for H1 ~ H6 are presented in (Table 3).

Table 3 Regression analysis results for hypothesis testing

Category			Unstandardized coefficients		Standardized Coefficients	t	Probability of significance
Hypothesis	Independent variable	Dependent variable	B	Std. error	Beta		
H1	(Constant)	Work engagement	0.882	0.285	–	3.091	0.002[b]
	Coaching behavior		0.739	0.056	0.647	13.162	0.000[a]
	$R^2 = 0.419$, $F = 173.235$[a], Durbin-Watson = 2.024						
H2	(Constant)	Job satisfaction	0.340	0.274	–	1.241	0.216
	Coaching behavior		0.864	0.054	0.719	16.032	0.000[a]
	$R^2 = 0.517$, $F = 257.013$[a], Durbin-Watson = 1.834						
H3	(Constant)	Quiet quitting	5.486	0.299	–	18.337	0.000[a]
	Coaching behavior		−0.468	0.059	−0.457	– 7.949	0.000[a]
	$R^2 = 0.208$, $F = 63.189$[a], Durbin-Watson = 1.947						
H4	(Constant)	Job satisfaction	0.894	0.201	–	4.444	0.000[a]
	Work engagement		0.820	0.043	0.779	19.265	0.000[a]
	$R^2 = 0.607$, $F = 371.123$[a], Durbin-Watson = 1.893						
H5	(Constant)	Quiet quitting	5.218	0.237	–	22.042	0.000[a]
	Work engagement		−0.451	0.050	−0.503	– 9.005	0.000[a]
	$R^2 = 0.253$, $F = 81.086$[a], Durbin-Watson = 2.108						
H6	(Constant)	Quiet quitting	5.191	0.228	–	22.808	0.000[a]
	Job satisfaction		– 0.438	0.047	−0.513	– 9.269	0.000[a]
	$R^2 = 0.264$, $F = 85.910$[a], Durbin-Watson = 2.079						

[a]$p < 0.001$

5.1 Mediation Effect Analysis

Analysis of the mediating variables' effects showed that work engagement significantly lowered quiet quitting levels ($\beta = -0.503$, $p < 0.001$). Job satisfaction was also found to have a significant negative (−) effect on quiet quitting ($\beta = -0.513$, $p < 0.001$). This indicates that higher work engagement and job satisfaction reduce the likelihood of members' quiet quitting. Furthermore, work engagement was analyzed to have a significant positive (+) effect on job satisfaction ($\beta = 0.779$, $p < 0.001$), showing a strong correlation between these two positive job attitudes. The hypothesis testing results related to the mediation effects H7 and H8 are presented in (Tables 4 and 5).

6 Conclusion

6.1 Research Results

The results regarding the influence relationships among the study variables are as follows. Work engagement was found to have a statistically significant negative (–) effect on members' levels of quiet quitting. Similarly, job satisfaction also showed a significant negative (–) impact on the tendency toward quiet quitting. These results suggest that higher levels of work engagement and job satisfaction are associated with lower possibilities of psychological withdrawal among members. Furthermore, work engagement demonstrated a significant positive (+) effect on job satisfaction, indicating a strong correlation between these two positive job attitude variables. The analysis of mediating variables revealed that work engagement significantly reduced the level of quiet quitting among members, exerting a negative (–) influence. In a similar vein, job satisfaction was also confirmed to have a significant negative (–) impact on the tendency toward quiet quitting. These findings indicate that the greater the members' work engagement and job satisfaction, the lower the likelihood of psychological withdrawal.

6.2 Academic Implications

This study makes an important contribution in empirically identifying the impact structure of leader coaching behavior on organizational members' quiet quitting. In particular, it has academic distinctiveness in comprehensively analyzing the complex mediating relationships between leader coaching behavior, quiet quitting, work engagement, and job satisfaction. By empirically revealing the existence of a mediating pathway where a leader's coaching behavior enhances members' work engagement and job satisfaction, thereby mitigating quiet quitting, this study enabled

Table 4 Leader coaching behavior and its influence on quiet quitting: mediating effect of work engagement

Category			Unstandardized coefficients		Standardized coefficients	t	Probability of significance	R^2	*F*
Stage	Independent variable	Dependent variable	B	Std. error	Beta				
1	(Constant)	Quiet quitting	5.486	0.299	–	18.337	0.000[a]	0.208	63.189[a]
	Coaching behavior		−0.468	0.059	−0.457	−7.949	0.000[a]		
2	(Constant)	Work engagement	0.882	0.285	–	3.091	0.002[b]	0.419	173.235[a]
	Coaching behavior		0.739	0.056	0.647	13.162	0.000[a]		
3	(Constant)	Quiet quitting	5.768	0.291	–	19.813	0.000[a]	0.282	46.970[a]
	Coaching behavior		−0.231	0.074	−0.226	−3.140	0.002[b]		
	Work engagement		−0.320	0.065	−0.356	−4.955	0.000[a]		

Table 5 Leader coaching behavior and its influence on quiet quitting: mediating effect of job satisfaction

Category			Unstandardized coefficients		Standardized coefficients	t	Probability of significance	R^2	*F*
Stage	Independent variable	Dependent variable	B	Std. error	Beta				
1	(Constant)	Quiet quitting	5.486	0.299	–	18.337	0.000[a]	0.208	63.189[a]
	Coaching behavior		0.468	0.059	−0.457	−7.949	0.000[a]		
2	(Constant)	Job satisfaction	0.340	0.274	–	1.241	0.216	0.517	257.013[a]
	Coaching behavior		0.864	0.054	0.719	16.032	0.000[a]		
3	(Constant)	Quiet quitting	5.597	0.287	–	19.504	0.000[a]	0.279	46.332[a]
	Coaching behavior		−0.185	0.081	−0.181	−2.288	0.023[c]		
	Job satisfaction		−0.327	0.067	−0.383	−4.852	0.000[a]		

an academic understanding of the mechanism of coaching behavior that leaders exercise on members [23]. This enriches academic discussion on the new phenomenon of quiet quitting and strengthens the theoretical basis for the effectiveness of positive coaching behavior in preventing members' psychological withdrawal [24].

6.3 Practical Implications

This study presents specific core strategies for effectively managing the phenomenon of organizational members' quiet quitting behavior. It was confirmed that a leader's coaching behavior directly reduces members' psychological withdrawal and indirectly mitigates quiet quitting by enhancing work engagement and job satisfaction. Therefore, organizations need systematic planning and implementation for developing and strengthening manager coaching competencies. This can be a key measure to effectively increase members' job commitment and satisfaction and prevent quiet quitting tendencies. Ultimately, this will bring practical effects of psychological well-being for members and performance improvement for organizations.

6.4 Limitations and Suggestions for Future Research

While this study empirically identified the relationships between coaching behavior, quiet quitting, work engagement, and job satisfaction, it has several limitations. First, due to the constraints of cross-sectional research design, it is difficult to definitively establish causal relationships between variables, necessitating longitudinal research to identify temporal effects. Second, multi-faceted data collection is suggested to complement the limitations of self-report surveys. Third, research is needed to expand results limited to specific samples to various industries and cultures. To complement the limitations of this study, it is necessary to clearly identify causal relationships between variables using a longitudinal research design and expand research results to various industries and samples in the future.

References

1. Harter J (2022) Is quiet quitting real? Gallup
2. Mahand T, Caldwell C (2023) Quiet quitting–causes and opportunities. Bus Manag Res 12(1):9–19
3. Zenger J, Folkman J (2022) The great resignation and quiet quitting paradigm shifts: an overview of current situation and future research directions. J Hosp Market Manag 31(8):899–907
4. Hunt JM, Weintraub A (2002) Coaching in organizations. Research Gate

5. Werner JM, Desimone RL (2006) Human resource development (4th ed.). South-Western College Pub
6. Cho EH, Tak JK (2011) The effects of transformational leadership on psychological empowerment and organizational commitment: the mediating effect of coaching behavior. Korean J Ind Organ Psychol 24(3):437–460
7. Kilburg RR (1996) Toward a conceptual understanding of coaching and executive coaching. Consult Psychol J: Pract Res 48(2):134
8. Whitmore J (2010) Coaching for performance: growing human potential and purpose: the principles and practice of coaching and leadership. Hachette UK
9. Lee YS, Cho DY, Kim JM (2018) The relationship between leaders' coaching behavior and members' innovative behavior: the mediating effect of empowerment. J Vocat Educ Res 37(2):163–186
10. Ellinger AD, Ellinger LM, Keller SB (2005) Supervisory coaching behavior, employee satisfaction, and warehouse employee performance: a structural model. Hum Resour Dev Q 16(3):343–368
11. Klotz AC, Bolino MC (2022) When quiet quitting is worse than the real thing. Harv Bus Rev Digit Artic 22(9):1–5
12. Oh HS (2024) Effect of job insecurity and performance compensation linkage on quiet resignation: mediating effect of psychological contract violation, moderating effect of individualism [Doctoral dissertation, Chosun University]
13. Wigert B (2022) Quiet firing: what it is and how to stop doing it. Workplace at Gallup
14. Schaufeli WB, Salanova M, González-Romá V, Bakker AB (2002) The measurement of engagement and burnout: a two-factor confirmatory analytic approach. J Happiness Stud 3(1):71–92
15. Bakker AB, Schaufeli WB, Leiter MP, Taris TW (2004) Work engagement: an emerging concept in occupational health psychology. Work Stress 18(2):170–188
16. Schaufeli WB, Bakker AB (2010) Job demands, job resources, and burnout: a longitudinal study. J Organ Behav 25(3):293–312
17. Bakker AB, Schaufeli WB, Leiter MP, Taris TW (2008) Work engagement: an emerging concept in occupational health psychology. Work Stress 22(3):187–202
18. Omarov A (2009) Organizational trust and job satisfaction [Unpublished master's thesis]. Dokuz Eylül University
19. Banerjee D, Perrucci CC (2010) Job satisfaction: Impact of gender, race, worker qcualificcatcions, and work context. In: Gender and sexuality in the workplace. Emerald Group Publishing Limited, pp 39–58
20. Skaalvik EM, Skaalvik S (2011) Teachers' feelings of belonging, emotional exhaustion, and job satisfaction: the importance of school goal structure and sense of belonging. Int J Educ Res 50(3):175–184
21. Bulińska-Stangrecka H, Bagieńska A (2021) The relationship between trust and job satisfaction of correctional officers. Int J Environ Res Public Health 18(13):6778
22. Jo Y, Hong G (2019) The effect of leader's coaching behavior on employees' work engagement: the mediating roles of self-efficacy and job meaningfulness, and the moderating role of trust. Korean J Manag 48(1):243–273
23. Ali M, Lodhi SA, Raza B, Ali W (2018) Examining the impact of managerial coaching on employee job performance: mediating role of work engagement, leader-member-exchange quality, job satisfaction, and turnover intentions. Pak J Commer Soc Sci 12(1):253–282
24. Lee MCC, Idris MA, Tuckey M (2018) Supervisory coaching and performance feedback as mediators of the relationships between leadership styles, work engagement, and turnover intention. Hum Resour Dev Int 22(1):1–26

A Study on the Influence of Work-Family Conflict on the Subjective Happiness of Dual-Income Women-Focusing on the Mediating Effects of Parenting Stress and Marital Satisfaction

I.-J. Shin and J.-w. Park

1 Introduction

Korean society currently faces a serious sustainability crisis. As of 2023, Korea's total fertility rate recorded the lowest among OECD countries at 0.72 children per woman, while the national happiness index stands at 6.15 points (out of 10), ranking 52nd among 143 countries (Statistics Korea 2023; World Happiness Report 2024). The fundamental causes of this low fertility rate and happiness index can be traced to structural issues such as gender discrimination, unbalanced distribution of household labor, and excessive childcare burden (Kang 2024; Kang and Kim 2023). Notably, in 48.2% of dual-income households, domestic duties and childcare are concentrated on women, and the resulting difficulties in balancing household and workplace roles generate negative emotions among working mothers (Judge et al. 2006). In this social context, this study focuses on dual-income women raising early adolescent children (ages 13–15) to analyze the impact of work-family conflict on subjective well-being. To enhance generalizability, we utilize data from the Panel Study on Korean Children and verify the mediating effects of parenting stress and marital satisfaction based on goal achievement theory (Im-Park and Oh 2012). From the perspectives of positive psychology and coaching psychology, this study aims to provide practical foundational data for establishing family-friendly policies and developing human resource management strategies for corporations in the ESG management era.

I.-J. Shin
Graduate School of Project Management, Soongsil University, Seoul, Korea
e-mail: injeong77@hanmail.net

J. Park (✉)
Department of Business Administration, Soongsil University, Seoul, Korea
e-mail: Jongpark7@ssu.ac.kr

R. Lee et al. (eds.), *IT Service & Business Innovation*, Studies in Computational Intelligence 1252, https://doi.org/10.1007/978-3-032-14397-6_10

2 Theoretical Background

2.1 *Work-Family Conflict*

Work-family conflict is defined as a form of inter-role conflict in which role pressures from work and family domains are mutually incompatible (Greenhaus and Beutell 1985). This conflict manifests in two major transfer phenomena: spillover, where stress from one domain transfers to another, and crossover, where stress is transferred between spouses. Specifically, husbands tend to show more prominent transfer of family stress to work, while wives tend to adjust their domestic labor in response to their spouse's work stress (Bolger et al. 1989).

Work-family conflict has been empirically proven to be closely related to individual mental health, as role conflicts between work and family can lead to problems such as depression, anxiety, and psychological strain (Grzywacz and Marks 2000).

2.2 *Work-Parenting Conflict*

Previous studies on work-parenting conflict have analyzed its multifaceted characteristics. Boundary theory explains the balance between work, family, and parenting domains (Clark 2000), while research has shown that time constraints due to work life induce feelings of guilt, and societal expectations regarding dual roles place psychological burden on working mothers (Arendell 2000). Therefore, work-family balance support requires an integrated approach that includes both physical and emotional support.

2.3 *Parenting Stress*

Previous studies have systematically analyzed the phenomenon of parenting stress. It has been defined as an imbalance between parental resources and parenting demands, and has been found to negatively affect mother–child relationships and family systems (Abidin 1992). While parenting stress among mothers with school-age children has been shown to impede family atmosphere and parent–child interactions, research has demonstrated that it can be managed through Achievement-focused Psychoeducational Therapy(APT) programs (Im-Park and Oh 2012).

2.4 *Marital Satisfaction*

Marital satisfaction has been defined as a subjective feeling and attitude regarding one's satisfaction and happiness with married life. It has evolved to encompass relationships with multiple factors including family affection, interpersonal relationships, economic stability, childbirth, and child-rearing. The concept has been further refined to include individual subjective feelings and attitudes toward married life, as well as the comparison between expectations and actual outcomes (Hawkins and Johnson 1969). The marital satisfaction scale incorporated specific elements such as expressions of affection and intimacy (Schumm et al. 1983). Notably, systematic research has been conducted on changes in wives' marital satisfaction during child-rearing periods and its impact on mother–child relationships (Fishman and Meyers 2000).

2.5 *Subjective Well-Being*

The academic definition and concept of subjective well-being has evolved over time. Happiness was initially defined as a comprehensive evaluation of life based on individual subjective experiences, and later developed into a positive assessment of one's internal state and life (Lyubomirsky and Lepper 1999). Research has revealed that it results from the interaction of various factors including internal characteristics, education, relationships, and health. More recently, it has come to be understood from a perspective that integrates both cognitive aspects and emotional experiences (Vaillant 2010).

3 Research Method

3.1 *Research Models*

This study aims to empirically analyze the impact of work-family conflict on subjective well-being among dual-income women, and to examine the significant mediating effects of parenting stress and marital satisfaction in this relationship. The research model for this analysis is presented in Fig. 1.

3.2 *Research Hypothesis*

This study established hypotheses based on the Conflict Model and Scarcity Hypothesis as theoretical foundations. According to the Conflict Model, work-family balance

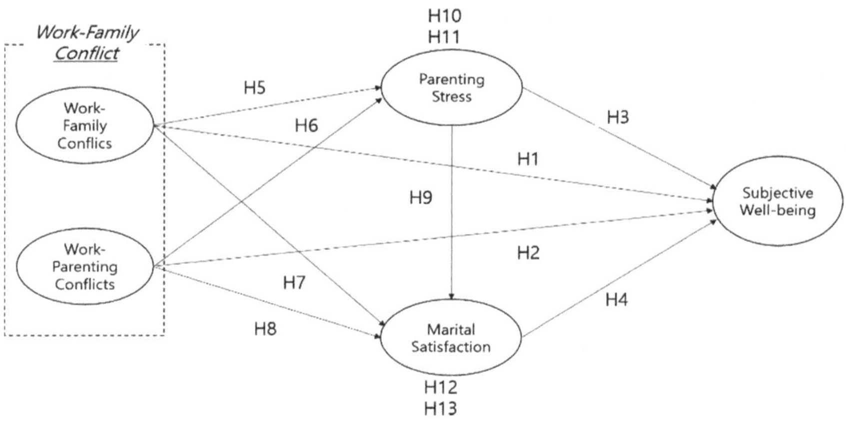

Fig. 1 Research model

is defined as a process of resolving conflicts arising from the collision between work and family roles (Greenhaus and Beutell 1985). Empirical studies have shown that working mothers' work-parenting conflict reduces both their subjective well-being and their children's life satisfaction (Han and Shin 2024). Based on these theoretical grounds and empirical evidence, hypotheses H1 and H2 were established.

H1: Work-family conflict will have a negative (−) effect on subjective well-being.

H2: Work-parenting conflict will have a negative (−) effect on subjective well-being.

According to previous studies, both parenting stress and marital satisfaction are closely related to subjective well-being. Abidin (1992) defined parenting stress as an imbalance between parental resources and parenting demands. Hawkins and Johnson (1969) reported that marital satisfaction affects the psychological health of all family members, and positive relationships between spouses have been shown to lead to improved psychological well-being and parenting behavior (Fishman and Meyers 2000). Based on these theoretical foundations, hypotheses H3 and H4 were established.

H3: Parenting stress will have a negative (−) effect on subjective well-being.

H4: Marital satisfaction will have a positive (+) effect on subjective well-being.

The hypotheses regarding the relationship between work-family conflict and parenting stress were established based on the following logical foundations. Work-family conflict, work-parenting conflict, and parenting stress refer to tension between work-family roles, work-parenting role conflict, and the burden of parental role performance, respectively, and these factors are closely interrelated (Im-Park and Oh 2012). Based on the findings of Shin and Cho (2023) that work-family conflict increases overall life stress and work-parenting conflict elevates parenting stress, this study established hypotheses H5 and H6, which propose that increased work-family conflict and work-parenting conflict will lead to increased parenting stress.

H5: Work-family conflict will have a positive (+) effect on parenting stress.

H6: Work-parenting conflict will have a positive (+) effect on parenting stress.

Previous studies have demonstrated the negative effects of work-family conflict, work-parenting conflict, and parenting stress on marital satisfaction from various angles. Son (2021) found that conflicts arising from performing multiple roles as an employee, spouse, and parent reduce marital satisfaction, with this tendency being particularly pronounced among women. Yoon (2024) confirmed that higher levels of parenting stress correlate with lower marital satisfaction. Based on this empirical evidence from previous studies, hypotheses H7, H8, and H9 were established, proposing that work-family conflict, work-parenting conflict, and parenting stress will negatively affect marital satisfaction.

H7: Work-family conflict will have a negative (−) effect on marital satisfaction.

H8: Work-parenting conflict will have a negative (−) effect on marital satisfaction.

H9: Parenting stress will have a negative (−) effect on marital satisfaction.

Numerous previous studies have demonstrated the mediating effects of parenting stress. Kang and Kim (2023) revealed that work-family role conflict increases parenting stress, which in turn decreases subjective well-being. Based on these findings, hypotheses H10 and H11 were established, proposing that parenting stress would mediate the relationship between work-family conflict and work-parenting conflict with subjective well-being.

H10: Parenting stress will mediate the relationship between work-family conflict and subjective well-being.

H11: Parenting stress will mediate the relationship between work-parenting conflict and subjective well-being.

The hypotheses regarding the mediating effects of marital satisfaction are based on the Actor-Partner Interdependence Model (APIM) theory and numerous empirical studies. Lee and Lee (2023) demonstrated that work-family conflict reduces life satisfaction through decreased marital satisfaction. Based on these previous studies, hypotheses H12 and H13 were established, proposing that marital satisfaction would mediate the relationship between work-family conflict and work-parenting conflict with subjective well-being.

H12: Marital satisfaction will mediate the relationship between work-family conflict and subjective well-being.

H13: Marital satisfaction will mediate the relationship between work-parenting conflict and subjective well-being.

3.3 Data Collection and Analysis Method

This study utilized data from the 14th wave (2021) of the Panel Study on Korean Children (PSKC) and conducted statistical analyses using IBM SPSS 22.0. The analysis included reliability and validity testing of measurement instruments, factor analysis, multiple regression analysis, and mediation effect analysis. The independent variables—work-family conflict and work-parenting conflict—were measured using a 5-point Likert scale, with reliable items selected through factor analysis. The dependent variable was subjective well-being, and the mediating variables were parenting stress and marital satisfaction, with mean values of each item used in the analysis. The PSKC, which was used in this study, is Korea's largest longitudinal study tracking children's growth and development processes, providing highly reliable data collected annually since 2008.

3.4 Measurement Instruments

The operational definitions of variables used in this study are presented in Table 1.

Table 1 Operational definitions of variables

Variable	Operational definition	Related research
Work-family conflict	Work-family conflict is defined as stress or role conflict arising from difficulties in balancing work and family life	Greenhaus and Beutell (1985)
Work-parenting conflict	Work-parenting conflict is defined as stress resulting from difficulties in simultaneously managing work and child-rearing responsibilities	Marshall and Barnett (1993)
Parenting stress	Parenting stress is defined as the burden of parental roles and perceived difficulties in raising children	Kim and Kang (1997)
Marital satisfaction	Marital satisfaction is defined as a subjective assessment that encompasses couples' overall satisfaction and happiness regarding their married life	Schumm et al. (1983)
Subjective well-being	Subjective well-being is defined as a comprehensive reflection of one's overall life happiness and satisfaction derived from comparisons with others	Lyubomirsky and Lepper (1999)

4 Research Results

4.1 Measurement Model Analysis

Based on the demographic characteristics analysis of the study participants (N = 630), all participants were mothers of panel children and resided with their children. The majority of mothers were in their early 40 s, with those aged 40–44 accounting for more than half (50.5%, n = 318). Regarding educational attainment, the largest proportion held a 4-year university degree (38.3%, n = 241), and the most common monthly household income range was 6–8 million KRW (31.6%, n = 199). In terms of employment status, the vast majority of mothers were currently employed (99.4%), while all fathers were employed. Most participants were in their first marriage (99.4%), and there was a relatively even distribution between male (52.7%) and female (47.3%) children.

4.2 Factor Analysis and Reliability

To verify the validity of measurement instruments, factor analysis was conducted using exploratory factor analysis (EFA). Specifically, principal component analysis with varimax rotation was applied. The KMO value was 0.927, and Bartlett's test of sphericity showed statistical significance ($p = 0.000$), confirming the appropriateness of factor analysis. Factor loadings ranged from 0.522 to 0.902, resulting in the extraction of two independent variables(work-family conflict and work-parenting conflict) and three mediating/dependent variables (parenting stress, marital satisfaction, and subjective well-being). Reliability analysis revealed that all measurement variables demonstrated satisfactory reliability with Cronbach's alpha coefficients exceeding 0.7. Specifically, the reliability coefficients were 0.910 for work-family conflict, 0.876 for work-parenting conflict, 0.891 for parenting stress, 0.930 for marital satisfaction, and 0.898 for subjective well-being.

4.3 Research Hypothesis Testing

4.3.1 Effects of Key Variables on Subjective Well-Being

Work-family balance conflict, parenting stress, and marital satisfaction all showed statistically significant relationships with subjective well-being. Analysis results revealed that work-family conflict, work-parenting conflict, and parenting stress had negative (−) effects on subjective well-being, while marital satisfaction showed a positive (+) effect. Specifically, higher levels of work-family conflict and parenting

Table 2 Effects of key variables on subjective well-being

Independent variables	Dependent variables	SE	β	t	p
Work-family conflict	Constant	0.113		51.324	0.000[1]
	Subjective well-being	0.051	−0.233	−6.002	0.000[1]
	Statistics	$R^2 = 0.054$, $\triangle R^2 = 0.053$, $F = 36.021$, $p = 0.000$[1]			
Work-parenting conflict	Constant	0.114		50.857	0.000[1]
	Subjective well-being	0.048	−0.236	−6.094	0.000[1]
	Statistics	$R^2 = 0.056$, $\triangle R^2 = 0.054$, $F = 37.137$, $p = 0.000$[1]			
Parenting stress	Constant	0.130		47.983	0.000[1]
	Subjective Well-being	0.047	−0.328	−8.693	0.000[1]
	Statistics	$R^2 = 0.107$, $\triangle R^2 = 0.106$, $F = 75.572$, $p = 0.000$[1]			
Marital satisfaction	Constant	0.161		17.268	0.000[1]
	Subjective well-being	0.043	0.514	15.010	0.000[1]
	Statistics	$R^2 = 0.264$, $\triangle R^2 = 0.263$, $F = 225.288$, $p = 0.000$[1]			

[1]$p < 0.001$

stress among working women were associated with lower levels of happiness, whereas higher marital satisfaction was associated with increased happiness.

All hypothesized relationships from H1 to H4 were statistically significant at the 95% confidence level. These findings suggest that to enhance subjective well-being, institutional support for work-family balance and practical assistance in reducing parenting stress are necessary. These analytical results are presented in Table 2.

4.3.2 Effects of Work-Family Balance Conflict on Parenting Stress and Marital Satisfaction

Testing the relationships between work-family balance conflict, parenting stress, and marital satisfaction variables revealed that all hypothesized relationships from H5 to H8 were statistically significant. Specifically, work-family balance conflict showed a positive (+) effect on parenting stress and a negative (−) effect on marital satisfaction. This indicates that as the imbalance between work and family life intensifies, the burden of child-rearing increases while the quality of marital relationships deteriorates. Furthermore, hypothesis H9, which proposed a negative (-) relationship between parenting stress and marital satisfaction, was also supported at a statistically significant level. This empirically demonstrates the causal relationship that higher parenting stress leads to lower marital satisfaction. These analytical results are presented in Table 3.

Table 3 Effects of work-family balance conflict on parenting stress and marital satisfaction

Independent variables	Dependent variables	SE	β	t	p
Work-family conflict	Constant	0.091		23.608	0.000[1]
	Parenting stress	0.041	0.241	6.210	0.000[1]
	Statistics	$R^2 = 0.058$, $\triangle R^2 = 0.056$, $F = 38.560$, $p = 0.000$[1]			
Work-parenting conflict	Constant	0.088		20.697	0.000[1]
	Parenting stress	0.037	0.384	10.435	0.000[1]
	Statistics	$R^2 = 0.148$, $\triangle R^2 = 0.146$, $F = 108.887$, $p = 0.000$[1]			
Work-family conflict	Constant	0.091		43.703	0.000[1]
	Marital satisfaction	0.041	−0.147	−3.721	0.000[1]
	Statistics	$R^2 = 0.022$, $\triangle R^2 = 0.020$, $F = 13.847$, $p = 0.000$[1]			
Work-parenting conflict	Constant	0.092		44.431	0.000[1]
	Marital satisfaction	0.039	−0.186	−4.756	0.000[1]
	Statistics	$R^2 = 0.035$, $\triangle R^2 = 0.033$, $F = 22.617$, $p = 0.000$[1]			
Parenting stress	Constant	0.107		39.285	0.000[1]
	Marital satisfaction	0.038	−0.205	−5.244	0.000[1]
	Statistics	$R^2 = 0.042$, $\triangle R^2 = 0.040$, $F = 27.497$, $p = 0.000$[1]			

[1] $p < 0.001$

4.3.3 Analysis Results of the Mediating Effects of Parenting Stress

This study analyzed the mediating effects of parenting stress between work-family balance and subjective well-being using Hayes' (2022) PROCESS macro Model 4. The results showed that both work-family conflict and work-parenting conflict had positive (+) effects on parenting stress. As these conflicts intensified, stress levels increased, leading to decreased subjective well-being.

Notably, the impact of work-parenting conflict was particularly prominent, confirming that work-parenting balance is crucial for enhancing the happiness of dual-income women. Parenting stress demonstrated significant mediating effects between work-family/work-parenting conflicts and subjective well-being (hypotheses H10, H11). The verification results are presented in Table 4.

Random resampling of 5,000 samples from the original data was conducted with a 95% confidence interval for analysis. The results revealed that both work-family conflict and work-parenting conflict had significant negative effects on subjective well-being through the mediation of parenting stress. Notably, conflicts related to child-rearing showed a greater impact on well-being compared to general family life conflicts. These analysis results are presented in Table 5.

Table 4 Mediating effects of parenting stress between work-family balance and subjective well-being

Path	*B*	*S.E*	*β*	*t*	*F*	R^2
Work-family conflict → Parenting stress	0.256	0.041	0.241	6.210[1]	38.560[1]	0.058
Work-family conflict → Subjective well-being	−0.215	0.050	−0.164	−4.268[1]	47.930[1]	0.133
Parenting stress → Subjective well-being	−0.356	0.047	−0.288	−7.526[1]		
Work-parenting conflict → Parenting stress	0.386	0.037	0.384	10.435[1]	108.887[1]	0.148
Work-parenting conflict → Subjective well-being	−0.160	0.050	−0.129	−3.192[1]	43.435[1]	0.122
Parenting stress → Subjective well-being	−0.343	0.050	−0.278	−6.857[1]		

[1] $p < 0.001$

Table 5 Bootstrapping results for the mediating effects of parenting stress

Independent variable	Mediator	Dependent variable	Effect	Boot. S.E	95% CI	
					Boot. LLCI	Boot. ULCI
Work-family conflict	Parenting stress	Subjective well-being	−0.091	0.021	−0.136	−0.052
Work-parenting conflict	Parenting stress	Subjective well-being	−0.132	0.025	−0.181	−0.086

4.3.4 Analysis Results of the Mediating Effects of Marital Satisfaction

The hypotheses H12 stating "Marital satisfaction has a mediating effect on the relationship between work-family conflict and subjective well-being" and H13 stating "Marital satisfaction has a mediating effect on the relationship between work-parenting conflict and subjective well-being" were both accepted at statistically significant levels. The research results showed that work-family conflict and work-parenting conflict had negative effects on both marital satisfaction and subjective well-being, while marital satisfaction had a positive effect on subjective well-being. Particularly, the finding that parenting-related conflicts had a greater impact on marital life suggests the importance of maintaining a balance between work, family, and parenting. The detailed statistical values and analysis results of these hypothesis tests are presented in Table 6.

The bootstrapping analysis results for the mediating effects of marital satisfaction showed that both work-family conflict and work-parenting conflict had negative effects on subjective well-being through the mediation of marital satisfaction.

Table 6 Mediating effects of marital satisfaction between work-family balance and subjective well-being

Path	*B*	*S.E*	*β*	*t*	*F*	R^2
Work-family conflict → Marital satisfaction	−0.153	0.041	−0.147	−3.721[1]	13.847[1]	0.022
Work-family conflict → Subjective well-being	−0.211	0.045	−0.161	−4.728[1]		
Marital satisfaction → Subjective well-being	0.617	0.043	0.490	14.403[1]	127.650[1]	0.289
Work-parenting conflict → Marital satisfaction	−0.183	0.039	−0.186	−4.756[1]	22.617[1]	0.035
Work-parenting conflict → Subjective well-being	−0.180	0.043	−0.146	−4.233[1]		
Marital satisfaction → Subjective well-being	0.613	0.043	0.487	14.154[1]	124.637[1]	0.284

[1] $p < 0.001$

Notably, work-parenting conflict demonstrated a greater impact on marital satisfaction and subjective well-being compared to work-family conflict. This suggests that in modern society, maintaining work-family balance, particularly managing parenting-related conflicts, is crucial, and that reducing such conflicts while improving marital satisfaction is necessary for enhancing individual well-being. These verification results are presented in Table 7.

Table 7 Bootstrapping results for the mediating effects of marital satisfaction

Independent variable	Mediator	Dependent variable	Effect	Boot. S.E	95% CI	
					Boot. LLCI	Boot. ULCI
Work-family conflict	Marital satisfaction	Subjective well-being	−0.095	0.029	−0.155	−0.040
Work-parenting conflict	Marital satisfaction	Subjective well-being	−0.112	0.026	−0.164	−0.064

5 Conclusion

5.1 Research Results

This study conducted model analysis on 630 dual-income women, and the main findings are as follows. First, work-family conflict, work-parenting conflict, and parenting stress had negative effects on subjective well-being, while marital satisfaction had a positive effect. Second, work-family This study conducted model analysis on 630 dual-income women, and the main findings are as follows. First, work-family conflict, work-parenting conflict, and parenting stress had negative effects on subjective well-being, while marital satisfaction had a positive effect. Second, work-family This study empirically analyzed factors affecting the subjective well-being of dual-income women. Specifically, we examined how work-family conflict and work-parenting conflict influence subjective well-being through the mediating effects of parenting stress and marital satisfaction, utilizing data from the 14th wave (2021) of the Panel Study on Korean Children. The main findings and implications derived from this empirical analysis are as follows. This study empirically analyzed factors affecting the subjective well-being of dual-income women. Specifically, we examined how work-family conflict and work-parenting conflict influence subjective well-being through the mediating effects of parenting stress and marital satisfaction, utilizing data from the 14th wave (2021) of the Panel Study on Korean Children. The main findings and implications derived from this empirical analysis are as follows. This study empirically analyzed factors affecting the subjective well-being of dual-income women. Specifically, we examined how work-family conflict and work-parenting conflict influence subjective well-being through the mediating effects of parenting stress and marital satisfaction, utilizing data from the 14th wave (2021) of the Panel Study on Korean Children. The main findings and implications derived from this empirical analysis are as follows. This study empirically analyzed factors affecting the subjective well-being of dual-income women. Specifically, we examined how work-family conflict and work-parenting conflict influence subjective well-being through the mediating effects of parenting stress and marital satisfaction, utilizing data from the 14th wave (2021) of the Panel Study on Korean Children. The main findings and implications derived from this empirical analysis are as follows.

To empirically analyze factors affecting the subjective well-being of dual-income women, this study conducted model analysis on 630 dual-income women using data from the 14th wave (2021) of the Panel Study on Korean Children. The analysis revealed the following results. First, work-family conflict, work-parenting conflict, and parenting stress had negative effects on subjective well-being, while marital satisfaction showed a positive effect. Second, work-family conflict and work-parenting conflict increased parenting stress and decreased marital satisfaction. Third, higher levels of parenting stress were associated with lower marital satisfaction. Fourth, both parenting stress and marital satisfaction were found to mediate the effects of work-family conflict and work-parenting conflict on subjective well-being. Finally, the difficulty of work-family balance was unrelated to age, and while higher education and income levels were associated with higher subjective well-being, the relationship between income and happiness did not show a strong correlation.

5.2 Academic Implications

The academic implications of this study are as follows. First, we conducted a detailed analysis of working women's conflict patterns by distinguishing work-family balance conflicts into work-family conflicts and work-parenting conflicts (Greenhaus and Beutell 1985). Second, we identified the pathway of decreased happiness among working women by verifying the sequential mediating effects of parenting stress and marital satisfaction (Kang and Kim 2023; Lee and Lee 2023). Third, we systematized parenting stress into three dimensions: daily stress, parental role burden, and the burden of delegating childcare to others (Kim and Kang 1997). Fourth, we empirically demonstrated the stress transfer process within the family system (Camisasca et al. 2014). Fifth, we verified the multidimensionality of happiness by integrating psychological well-being models and positive psychology (Keyes et al. 2002).

5.3 Practical Implication

The practical implications of this study are as follows. First, integrated support measures such as flexible work arrangements, parental leave, and workplace childcare facilities are necessary to alleviate work-family conflict (Kossek et al. 2011). Second, active participation of spouses in household childcare and domestic duties is essential, highlighting the importance of promoting paternity leave (Seymour et al. 2021). Third, organizations should implement managerial training and improve evaluation systems to foster family-friendly organizational cultures (Fiksenbaum 2014). Fourth, findings indicate simultaneous improvements in both working women's multiple role performance compensation and organizational performance, suggesting that work-family balance policies positively impact both organizations and individuals (Lim and Cho 2024).

5.4 Limitations and Future Research Directions

This study has several limitations that suggest directions for future research. First, due to the cross-sectional research design, we were unable to capture temporal changes in variables, and the panel data had limited sub-variables. Future research should analyze changes in work-family conflict and subjective well-being through longitudinal studies, develop multidimensional measurement tools for work-family conflict, and derive sustainable work-family balance culture formation measures through comprehensive analysis based on person-environment fit theory and work-family boundary theory (Clark 2000; Kossek et al. 2011).

References

Abidin RP (1992) The determinants of parenting behavior. J Clin Child 21(4):401–412

Arendell T (2000) Conceiving and investigating motherhood: the decade's scholarship. J Marriage Fam 62(4):1192–1207

Bolger N, DeLongis A, Kessler RC, Wethington E (1989) The contagion of stress across multiple roles. J Marriage Fam 51(1):175–183

Camisasca E, Miragoli S, Di Blasio P (2014) Is the relationship between marital adjustment and parenting stress mediated or moderated by parenting alliance? Eur J Psychol 10(2):235–254

Clark S (2000) Work-family border theory: a new theory of work-family balance. Hum RelatS 53(6):747–770

Fiksenbaum LM (2014) Supportive work–family environments: implications for work–family conflict and well-being. Int J Hum Resour Manag 25(5):653–672

Fishman EA, Meyers SA (2000) Marital satisfaction and child adjustment: direct and mediated pathways. Contemp Fam Ther 22(4):437–452

Greenhaus JH, Beutell NJ (1985) Sources of conflict between work and family roles. Acad Manag Rev 10(1):76–88

Grzywacz J, Marks N (2000) Reconceptualizing the work–family interface: an ecological perspective on the correlates of positive and negative spillover between work and family. J Occup Health Psychol 5(1):111–126

Han S, Shin N (2024) The effects of working mothers' work-family conflict and benefits on adolescents' life satisfaction: the mediating effect of working mothers' subjective well-being. In: Korean home economics association Spring conference proceedings, pp 505–506

Hawkins JL, Johnsen K (1969) Perception of behavioral conformity, imputation of consensus, and marital satisfaction. J Marriage Fam 31(3):507–511

Hayes AF (2022) Introduction to mediation, moderation, and conditional process analysis: a regression-based approach, 3rd edn. The Guilford Press, New York, NY

Im Park K, Oh S (2012) Effects of active parenting today based on goal attainment theory on parenting stress, parenting behavior, and parenting satisfaction in mothers of school-age children. J Korean Acad Nurs 42(5):659–670

Judge TA, Ilies R, Scott BA (2006) Work–family conflict and emotions: effects at work and at home. Pers Psychol 59(4):779–814

Kang B, Kim K (2023) The moderating effect of fathers' parenting involvement on the relationship between mothers' parenting stress and happiness for mothers with young children. Early Child Educ Res Rev 27(2):75–97

Kang E (2024) The effects of women's parenting stress on parenting efficacy: dual mediating effects of mental health and hope, and conditional indirect effects of employment status. Doctoral dissertation, Hanseo University

Keyes CLM, Shmotkin D, Ryff CD (2002) Optimizing well-being: the empirical encounter of two traditions. J Pers Soc Psychol 82(6):1007–1022

Kim K, Kang H (1997) Development of the parenting stress scale. J Korean Home Econ Assoc 35(5):141–150

Korea Institute of Child Care and Education (2022) Panel Study of Korean Children 14th Wave (2021) Data. https://panel.kicce.re.kr/

Kossek EE, Pichler S, Bodner T, Hammer LB (2011) Workplace social support and work–family conflict: a meta-analysis clarifying the influence of general and work–family-specific supervisor and organizational support. Pers Psychol 64(2):289–313

Lee S, Lee U (2023) The effects of working mothers' proactive personality on parenting behavior: the mediating effects of work-family conflict and marital satisfaction. Korean J Child Stud 44(3):251–264

Lim JH, Cho OS (2024) Effect of female managers' family-friendly system, supervisors' procedural fairness, and benefits on work-family balance. J Ind Converg 22(8):95–103

Lyubomirsky S, Lepper H (1999) A measure of subjective happiness: preliminary reliability and construct validation. Soc Indic Res 46(2):137–155

Marshall ML, Barnett RC (1993) Work-family strains and gains among two-earner couples. J Community Psychol 21(1):64–78

NPJ Data Analysis Institute (2021) Thesis Writing Using SPSS and PROCESS MACRO. Seoul: Good Land

Panel Study of Korean Children (2024) Tool profile. https://panel.kicce.re.kr/pskc/module/rawDataManage/index.do?menu_idx=56. Accessed 5 Jan 2024

Schumm WR, Nichols CW, Schectman KL, Grigsby CC (1983) Survey comparability of several self-report measures of marital instability. Psychol Rep 53(2):329–330

Seymour M, Peace R, Wood CE, Jillard C, Evans K, O'Brien J, Williams LA, Brown S, Giallo R (2021) We're in the background: facilitators and barriers to fathers' engagement and participation in a health intervention during the early parenting period. Health Promot J Aust 32:78–86

Shin I, Cho D (2023) The effect of work-family balance on life stress among working mothers with children: the mediating effect of marital intimacy. Korean J Early Child Educ 43(6):537–560

Son J (2021) Gender role attitudes, work-parenting conflict, and marital satisfaction among dual-earner couples with preschool children. Korean J Early Child Educ 44(2):23–55

Statistics Korea (2023) Population trends survey. http://kostat.go.kr

Vailant GE (2010) Aging well. New York, NY: Little, Brown and Company (Original work published 2002)

World Happiness Report (2024) World Happiness Report 2024. United Nations Sustainable Development Solutions Network

Yoon J (2024) Analysis of the relationship between spouse's smartphone overdependence, parenting stress, and marital satisfaction: focusing on parents with infants. In: Korean panel study of early childhood education and care conference, vol 2, pp 130–133

The Impact of Emotional Intelligence on Career Adaptability in Adolescents in the Age of AI

So-Im Park and Jong-Woo Park

1 Introduction

Historically, the advancement of human society and technology has continuously redefined the types and levels of core competencies required of adolescents in each era. The rapidly changing job market of the twenty-first century, driven by AI technology, demands self-directed adaptability to fluid career environments from adolescents (Pinski and Benlian 2024), highlighting the importance of comprehensive competencies for continuous career development beyond mere job preparation.

The normalization of AI technology has brought emotional communication, empathy, collaboration, self-direction, and critical understanding to the forefront as essential emotional and social cognitive competencies that are difficult for AI to replace (Pinski and Benlian 2024). In this context, emotional intelligence is emphasized as a key psychological and social resource for enhancing adolescents' career adaptability. Emotional intelligence is defined as the ability to recognize, understand, control, and utilize one's own emotions, playing a crucial role in forming emotional stability, interpersonal skills, and resilience even in uncertain environments (Halimi and AlShammari 2021; Mittal 2021; Mayang 2025).

This study aims to deeply investigate the impact of emotional intelligence on career adaptability, setting AI literacy as a social cognitive resource and growth mindset as a psychological resource as mediating variables. AI literacy provides the cognitive foundation necessary for adolescents' career exploration and design through the ability to critically understand and evaluate the principles, limitations, and ethical implications of AI (Carolus et al. 2023; Grassini 2024; Marienko et al.

S.-I. Park
Graduate School of Project Management, Soongsil University, Seoul, Korea
e-mail: rudghk102@hanmail.net

J.-W. Park (✉)
Department of Business Administration, Soongsil University, Seoul, Korea
e-mail: Jongpark7@ssu.ac.kr

R. Lee et al. (eds.), *IT Service & Business Innovation*, Studies in Computational Intelligence 1252, https://doi.org/10.1007/978-3-032-14397-6_11

2024), while a growth mindset refers to the belief that intelligence and abilities can be improved through effort and learning, helping to strengthen resilience by perceiving change and challenges as learning opportunities (Burnette et al. 2020; Yeager and Dweck 2020).

Previous studies have reported that emotional intelligence, AI literacy, and growth mindset each positively influence adolescents' career development (Halimi and AlShammari 2021; Mills et al. 2024). However, research analyzing the integrated relationships among these three variables structurally is still insufficient. In particular, empirical exploration of the mechanisms through which growth mindset and AI literacy mediate the impact of emotional intelligence on career adaptability is lacking.

Therefore, this study aims to elucidate the complex influence pathways of these three competencies on career adaptability, thereby expanding existing research and contributing to the establishment of career education directions for adolescents in the AI era.

2 Theoretical Background

2.1 Emotional Intelligence

Emotional intelligence is defined as the ability to recognize, understand, control, and utilize one's own and others' emotions and integrate them into thought and behavior (Bru-Luna et al. 2021). Salovey and Sluyter (1997) describe it through four subcomponents: emotional perception, utilization, understanding, and regulation. Husain et al. (2022) emphasize that emotional intelligence is closely related to both internal and external emotional satisfaction, influencing an individual's psychological well-being and performance. Emotional intelligence has been proven to positively impact academic performance (Halimi and AlShammari 2021), job satisfaction and performance (Wiguna et al. 2020), and interpersonal relationships and social adaptability (Mayang 2025). Particularly in the AI era, the unique emotional capabilities of humans emerge as an irreplaceable competitive advantage, making the development of emotional intelligence in adolescents an essential competency for future career development and successful social adaptation.

2.2 Growth Mindset

The growth mindset, based on Dweck's (2006) theory, refers to the belief that intelligence and abilities can be enhanced through effort and learning. This contrasts with a fixed mindset, which holds that abilities are static. Individuals with a growth mindset tend to view failures as opportunities for growth and strengthen their resilience through challenges and learning (Burnette et al. 2020; Park et al. 2020), thereby enhancing their psychological capacity for self-efficacy (Wiguna et al. 2020). This psychological attitude positively influences various areas, including academic performance, long-term career development, self-direction, and entrepreneurship (Burnette et al. 2020). In the rapidly changing AI-based society, a growth mindset is recognized as a key competency that provides a foundation for continuous learning and self-renewal, enabling flexible adaptation and proactive career planning in unpredictable career environments.

2.3 AI Literacy

AI literacy refers to a comprehensive ability to understand the basic principles of AI technology, critically analyze the results created by AI, and utilize the technology in a responsible manner (Mills et al. 2024; Pinski and Benlian 2024). Alamäki et al. (2024) emphasize that AI literacy should encompass not only technical understanding but also critical reflection on the ethical and social implications of AI. The main components include understanding AI concepts and principles, the ability to utilize AI technologies, social responsibility and ethical considerations regarding AI, and the ability to critically evaluate AI-based information (Carolus et al. 2023; Marienko et al. 2024). Fostering AI literacy in adolescents is essential for developing social competencies that go beyond mere technical use, enabling them to grow into active and ethical citizens in a rapidly changing digital society, effectively collaborating with AI, and adapting flexibly to changing job environments (Jo and Jung 2024).

2.4 Career Adaptability

Career adaptability refers to the ability to respond flexibly to changing job environments and to actively plan and manage one's career (Savickas and Porfeli 2012; Johnston 2018). Savickas and Porfeli (2012) describe it through four subcomponents: concern, control, curiosity, and confidence. Career adaptability serves as an important psychosocial resource for career development from adolescence to adulthood (Delle and Searle 2022; Lee and Kang 2025), with individuals possessing high career adaptability tending to explore new opportunities and respond flexibly to unexpected situations (Lee and Choi 2022; Lee and Kang 2025). In modern society, where job structures are rapidly reshaped by the widespread adoption of AI technology, career adaptability is increasingly recognized as a core competency that enables self-directed career development and lifelong learning (Maggiori et al. 2017).

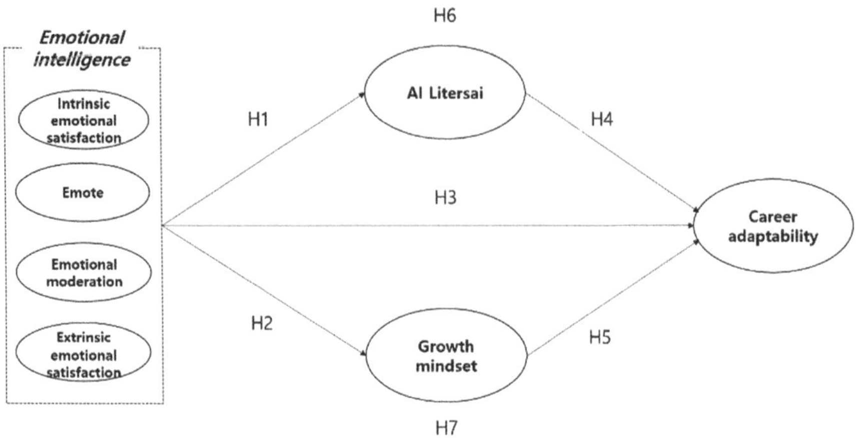

Fig. 1 Research model

3 Research Design

3.1 Research Model

This study aims to analyze the impact of emotional intelligence on career adaptability in adolescents living in the AI era and to empirically examine the structure in which AI literacy and growth mindset mediate this process. To achieve this, a systematic review of relevant prior research was conducted to establish the theoretical background, based on which the research model shown in Fig. 1 was derived.

3.2 Setting Research Hypotheses

Based on prior research findings that adolescents with high emotional intelligence are more likely to adapt flexibly to technological changes and have higher intrinsic motivation for learning new information, the following hypothesis was established (Husain et al. 2022; Pinski and Benlian 2024).

H1. Emotional intelligence will have a positive (+) effect on AI literacy.

Additionally, emotional intelligence contributes to accepting challenging situations and transforming failures into learning opportunities through self-understanding and emotional regulation. Based on research findings, the following hypothesis was proposed (Blackwell et al. 2007; Mittal 2021; Yeager and Dweck 2020).

H2. Emotional intelligence will have a positive (+) effect on a growth mindset.

Emotional intelligence provides flexibility and resilience in the career exploration process based on stable emotional management and interpersonal skills, which can contribute to enhancing career adaptability. Therefore, the following hypothesis was established based on the theoretical background (Savickas and Porfeli 2012; Halimi and AlShammari 2021; Maggiori et al. 2017).

H3. Emotional intelligence will have a positive (+) effect on career adaptability.

The advancement of AI technology has brought about changes across the entire job structure, making understanding and utilizing AI a core element of career adaptability (Mills et al. 2024). In fact, based on research findings that adolescents with high AI literacy exhibit higher confidence and proactive attitudes in career choices and transitions, the following hypothesis was derived (Grassini 2024).

H4. AI literacy will have a positive (+) effect on career adaptability.

A growth mindset, based on a positive perception of change and challenges, can directly contribute to enhancing career adaptability (Dweck 2006; Burnette et al. 2020; Wiguna et al. 2020; Park et al. 2020). Therefore, the following hypothesis was established.

H5. A growth mindset will have a positive (+) effect on career adaptability.

Emotional intelligence can stimulate motivation for learning about AI, which may act as an indirect positive influence on career adaptability through the enhancement of AI literacy. Based on this theoretical basis, the following mediation hypothesis was proposed (Pinski and Benlian 2024; Marienko et al. 2024).

H6. AI literacy will mediate the relationship between emotional intelligence and career adaptability.

Finally, based on existing research that suggests adolescents with high emotional intelligence recognize the growth potential in their abilities and exhibit positive attitudes in challenging situations, the following hypothesis was established, indicating that growth mindset mediates the relationship between emotional intelligence and career adaptability (Burnette et al. 2020; Wiguna et al. 2020; Dweck 2006).

H7. Growth mindset will mediate the relationship between emotional intelligence and career adaptability.

3.3 Measurement Tools

In this study, validated existing scales were used to measure emotional intelligence, AI literacy, growth mindset, and career adaptability. Emotional intelligence was assessed using a 12-item scale developed by Husain et al. (2022), evaluating emotional expression, emotional regulation, intrinsic emotional satisfaction, and extrinsic emotional satisfaction on a 7-point Likert scale. The growth mindset was measured using a 6-item scale based on Blackwell et al. (2007) and assessed on a 6-point Likert scale. AI literacy was measured using the PAILQ-6 developed by Grassini (2024), assessing cognitive usefulness and ethical sensitivity through 6 items on a 7-point Likert scale. Career adaptability was measured using one representative item from each sub-factor of the super-short form of the Career Adapt-Abilities Scale (CAAS-SSF) developed by Wang et al. (2024), consisting of a total of 4 items. Each

Table 1 Demographic characteristics

Variable	Item	Frequency	Percentage (%)
Gender	Male	175	57.0
	Female	132	43.0
Grade	Middle School 1st year	35	11.4
	Middle School 2nd year	53	17.3
	Middle School 3rd year	61	19.9
	High School 1st year	74	24.1
	High School 2nd year	46	15.0
	High School 3rd year	38	12.4
Region	Capital area	146	47.6
	Metropolitan city	130	42.3
	Small/Mid-sized city	31	10.1

item was measured on a 5-point Likert scale to assess the level of agreement of the respondents.

4 Research Analysis Results

4.1 Data Collection and Sample Characteristics

This study aimed to empirically analyze the relationships among emotional intelligence, AI literacy, growth mindset, and career adaptability of adolescents living in the AI era through a survey. Data collection was conducted from May 26 to May 31, 2025. The survey targeted adolescents who had experience utilizing AI in their academic work, and out of 350 surveys collected, 307 valid responses (87.7%) were used for final analysis. The demographic characteristics of the participating adolescents are presented in Table 1.

4.2 Validity and Reliability Verification

An exploratory factor analysis (EFA) was conducted to ensure the construct validity of the measurement tools for emotional intelligence, AI literacy, growth mindset, and career adaptability. The analysis utilized principal axis factoring and varimax rotation, confirming the suitability of factor analysis through the Kaiser-Meyer-Olkin (KMO) measure of sampling adequacy and Bartlett's test of sphericity.

The KMO measure was found to be 0.904, indicating an excellent level, and the approximate chi-square value from Bartlett's test was significant at 4434.395 ($p < 0.001$), confirming that the prerequisites for factor analysis were met.

Based on the criteria of eigenvalues greater than 1 and the scree plot, a total of 7 factors were extracted, with a cumulative explanatory power of 69.477%. This indicates that each measurement variable was appropriately separated into the theoretically established 7 factors (intrinsic emotional satisfaction, emotional expression, emotional regulation, extrinsic emotional satisfaction, AI literacy, growth mindset, career adaptability), and all items were clearly loaded onto their respective factors. The specific factor loadings ranged from 0.688 to 0.832, ensuring construct validity through consistent cohesion with theoretical variables and minimizing cross-loadings.

The reliability analysis of the main variables showed that all scales met the reliability verification criteria. The Cronbach's α coefficients of each item were as follows: intrinsic emotional satisfaction (0.740), emotional expression (0.796), emotional regulation (0.849), extrinsic emotional satisfaction (0.828), AI literacy (0.883), growth mindset (0.899), and career adaptability (0.847), all exceeding 0.70, indicating sufficient internal consistency among the measurement items. Therefore, the measurement items utilized in this study are deemed statistically reliable.

4.3 Hypothesis Testing

H1. The Effect of Emotional Intelligence on AI Literacy

The results of the multiple regression analysis indicated that emotional intelligence significantly influenced AI literacy. Notably, extrinsic emotional satisfaction had the most substantial positive impact on the level of AI literacy, suggesting the importance of emotional interactions as social cognitive resources in developing AI literacy (Table 2).

H2. The Effect of Emotional Intelligence on Growth Mindset

The analysis of the effect of emotional intelligence on growth mindset revealed that the regression model is statistically significant. Intrinsic emotional satisfaction had the strongest positive effect on growth mindset, indicating that the ability to accept and control emotions contributes to enhancing perceptions of personal growth potential (Table 2).

H3. The Effect of Emotional Intelligence on Career Adaptability

The analysis of the effect of emotional intelligence on career adaptability showed that the overall regression model was statistically significant. Emotional expression had the most substantial positive impact on career adaptability, suggesting that career adaptability is closely related not only to cognitive abilities but also to emotional competencies (Table 2).

H4 and 5 The Effects of AI Literacy and Growth Mindset on Career Adaptability

The analysis of the effects of AI literacy and growth mindset on career adaptability indicated that the overall model was significant. AI literacy showed the most substantial positive effect on career adaptability, while growth mindset also had a significant positive impact. This empirically demonstrates the importance of AI utilization skills and positive self-belief in the development of adolescents' career adaptability (Table 2).

4.4 Mediation Effect Analysis

H6. Mediation Effect of AI Literacy

To verify the mediation effect of AI literacy on the influence of emotional intelligence on career adaptability, a three-step regression analysis was conducted based on Baron and Kenny's (1986) method. The analysis revealed that in the first step, emotional expression, emotional regulation, and extrinsic emotional satisfaction significantly influenced AI literacy. In both the second and third steps, the results were statistically significant. Notably, the significant direct effect of emotional intelligence on career adaptability identified in the second step decreased when AI literacy was included in the third step, demonstrating a significant partial mediation effect of AI literacy. These results empirically support the notion that social cognitive resources, such as AI literacy, play a crucial mediating role in the process through which emotional intelligence affects adolescents' career adaptability (Table 3).

H7. Mediation Effect of Growth Mindset

To verify the mediation effect of growth mindset in the relationship between emotional intelligence and career adaptability, a three-step regression analysis was conducted based on Baron and Kenny's (1986) method. The analysis revealed that all steps yielded statistically significant t-values and p-values ($p < 0.001$). Notably, the significant direct effect of emotional intelligence on career adaptability identified in the second step decreased when growth mindset was included in the third step, demonstrating a significant partial mediation effect of growth mindset. These results indicate that a growth mindset serves as a key mediating pathway in the process through which emotional intelligence influences adolescents' career adaptability. In other words, adolescents' emotional intelligence contributes to enhancing their adaptability to unpredictable career environments by strengthening their growth mindset (Table 4).

5 Conclusion and Suggestions

This study aimed to empirically verify the structural relationships among emotional intelligence, AI literacy, growth mindset, and career adaptability among adolescents living in the AI era, conducting multiple regression and mediation effect analyses with 307 adolescents.

Table 2 Results of regression analysis for hypotheses 1, 2, 3, 4, and 5

H	IV	DV	Unstandardized			*t*	Sig
			B	S.E	β		
H1-1	Intrinsic emotional satisfaction	AI literacy	0.049	0.051	0.052	0.969	0.333
H1-2	Emotional expression		0.225	0.053	0.243	4.278	0.000
H1-3	Emotional regulation		0.135	0.046	0.159	2.933	0.004
H1-4	Extrinsic emotional satisfaction		0.235	0.048	0.268	4.872	0.000
R^2: 0.291, *F*: 30.927, *p:* 0.000, Durbin-Watson: 1.845,							
H2-1	Intrinsic emotional satisfaction	Growth mindset	0.270	0.055	0.264	4.862	0.000
H2-2	Emotional expression		0.186	0.057	0.186	3.236	0.001
H2-3	Emotional regulation		0.099	0.050	0.108	1.967	0.050
H2-4	Extrinsic emotional satisfaction		0.152	0.053	0.161	2.898	0.004
R^2: 0.272, *F*: 28.191 *p:* 0.000, Durbin-Watson: 2.051							
H3-1	Intrinsic emotional satisfaction	Career adaptability	0.222	0.051	0.217	4.325	0.000
H3-2	Emotional expression		0.269	0.053	0.269	5.068	0.000
H3-3	Emotional regulation		0.147	0.046	0.160	3.168	0.002
H3-4	Extrinsic emotional satisfaction		0.197	0.049	0.208	4.046	0.000
R^2: 0.379, *F*: 46.097 *p:* 0.000, Durbin-Watson: 1.972							
H4	AI literacy	Career adaptability	0.408	0.054	0.378	7.611	0.000
H5	Growth mindset		0.314	0.050	0.314	6.316	0.000
R^2: 0.379, *F*: 46.097 *p:* 0.000, Durbin-Watson: 1.972							

Table 3 Results of mediation effect analysis for Hypothesis 6

MEVS	IV	DV	B	β	*t*	Sig
1st step	Intrinsic emotional satisfaction	AI literacy	0.049	0.052	0.969	0.333
	Emotional expression		0.225	0.243	4.278	0.000
	Emotional regulation		0.135	0.159	2.933	0.004
	Extrinsic emotional satisfaction		0.235	0.268	4.872	0.000
	R^2: 0.291, *F*: 30.927, *p*: 0.000, Durbin-Watson: 1.845					
2nd step	Intrinsic emotional satisfaction	Career adaptability	0.222	0.217	4.325	0.000
	Emotional expression		0.269	0.269	5.068	0.000
	Emotional regulation		0.147	0.160	3.168	0.002
	Extrinsic emotional satisfaction		0.197	0.208	4.046	0.000
	R^2: 0.379, *F*: 46.097, *p*: 0.000, Durbin-Watson: 1.972					
3rd step	Intrinsic emotional satisfaction	Career adaptability	0.211	0.206	4.202	0.000
	Emotional expression		0.219	0.219	4.098	0.000
	Emotional regulation		0.117	0.128	2.544	0.011
	Extrinsic emotional satisfaction		0.144	0.153	2.927	0.004
	AI literacy		0.223	0.207	3.928	0.000
	R^2: 0.409, *F*: 41.726, p: 0.000, Durbin-Watson: 2.025					

5.1 Summary of Research Results

The main analysis results of this study are as follows:

First, emotional intelligence had a significant positive effect on AI literacy, with extrinsic emotional satisfaction being crucial for developing AI literacy (Pinski and Benlian 2024; Marienko et al. 2024). This highlights the importance of positive emotional experiences when utilizing AI.

Table 4 Results of mediation effect analysis for Hypothesis 7

MEVS	IV	DV	B	β	*t*	Sig
1st step	Intrinsic emotional satisfaction	Growth mindset	0.270	0.264	4.862	0.000
	Emotional expression		0.186	0.186	3.236	0.001
	Emotional regulation		0.099	0.108	1.967	0.050
	Extrinsic emotional satisfaction		0.152	0.161	2.898	0.004
	R^2: 0.272, *F*: 28.191, *p:* 0.000, Durbin-Watson: 2.051					
2nd step	Intrinsic emotional satisfaction	Career adaptability	0.222	0.217	4.325	0.000
	Emotional expression		0.269	0.269	5.068	0.000
	Emotional regulation		0.147	0.160	3.168	0.002
	Extrinsic emotional satisfaction		0.197	0.208	4.046	0.000
	R^2: 0.379, *F*: 46.097, *p:* 0.000, Durbin-Watson: 1.972					
3rd step	Intrinsic emotional satisfaction	Career adaptability	0.183	0.179	3.474	0.001
	Emotional expression		0.242	0.242	4.536	0.000
	Emotional regulation		0.133	0.145	2.874	0.004
	Extrinsic emotional satisfaction		0.175	0.185	3.584	0.000
	Growth mindset		0.144	0.144	2.730	0.007
	R^2: 0.394, *F*: 39.156, p: 0.000, Durbin-Watson: 1.918					

Second, emotional intelligence had a significant positive effect on growth mindset, with intrinsic emotional satisfaction being important for forming positive beliefs (Blackwell et al. 2007; Dweck 2006). This indicates that the acceptance and regulation of emotions enhance beliefs in one's potential for growth.

Third, emotional intelligence had a significant positive effect on career adaptability, with emotional expression being crucial for adapting to uncertain career situations (Savickas and Porfeli 2012; Halimi and AlShammari 2021). This suggests

2 Theoretical Background

2.1 Social Capital

Social capital refers to the capacity for trust, communication, and cooperation within social structures [6], and is conceptualized as a collective of resources accessible through social relationships [7]. As an intangible capital, social capital plays a crucial role in facilitating collective action for achieving common goals and solving problems, based on trust, norms, and networks [8]. Through studies by Coleman [9], Bourdieu [10], Putnam [11], and others, social capital has gained international recognition as a solution to weakening social cohesion. In this study, trust, norms, and networks are established as sub-variables of social capital.

2.1.1 Trust

Trust is a fundamental element of social capital and serves as a key mechanism for strengthening bonds between individuals and groups. It is defined as psychological security gained through expectations of others' trustworthiness [12], anticipated cooperative behavior based on community norms [2], or the subjective probability that others will act as expected [13]. Trust is categorized into interpersonal trust targeting specific individuals and institutional trust targeting social systems, encompassing reliability, predictability, understanding, and interdependence [14]. It is formed through fairness, communication, and authenticity, playing important roles in apartment communities such as strengthening resident cohesion, improving community operational efficiency, and reducing conflicts [15].

2.1.2 Norms

Norms are essential components of social capital, serving as informal behavioral guidelines that maintain social stability and promote cooperation for achieving common goals [9]. They foster reciprocity to regulate self-interested behavior and mitigate 'free-rider problems,' while establishing mutual obligations and expectations to inhibit opportunistic behavior [16]. Norms are considered core elements including reciprocity and regulatory functions [17], emphasizing the importance of generalized reciprocity norms for social efficiency [11]. They are classified into limited and generalized reciprocity based on the scope of mutual exchange, affecting individual satisfaction and group problem-solving capabilities [18]. In residential communities, norms contribute to quality of life improvement by regulating daily dynamics such as shared resource usage and noise levels, thereby strengthening community cohesion and interpersonal trust.

2.1.3 Network

Networks represent the structural aspect of social capital, forming relationship webs through social interaction. While evolving with digitalization, they serve as crucial pathways for social capital formation and resource access, encompassing systematic building processes beyond mere voluntary formation [19]. Social networks were presented as concepts including all personal relationships [20], with Coleman [17] emphasizing networks as essential information channels and core components of social capital. Putnam [11] viewed networks as the foundation of social capital, arguing that horizontal networks are effective for social integration. They develop through trust-based relationship systems and manifest in various forms such as bonding, bridging, and linking networks based on tie strength [2]. Informal networks tend to promote specific trust patterns, formal networks foster generalized trust, and the importance of linking networks connecting various social strata is also emphasized [21].

2.2 Sense of Community

Sense of community refers to the psychological bond where individuals feel belonging to a specific community and desire to contribute to its development. It serves as a basic psychological structure for individuals to discover identity and form social relationships, encompassing both geographical and relational dimensions and characterized by common goals, norms, and group activities [22]. Sense of community is recognized as a key element in solving modern social problems, with Sarason [22] emphasizing various definitions including interdependence and obligation. The most widely used framework for understanding community sense consists of four core components proposed by McMillan and Chavis [23]. Membership, influence, integration and fulfillment of needs, and shared emotional connection include feelings of belonging to the community, perception of influencing and being influenced by the community, awareness of fulfilling needs and acquiring resources through the community [24], and sense of emotional connection with community members. Sense of community is formed through interaction between individuals and community structures, serving as an important factor promoting community participation and positively impacting individual and community well-being [22].

2.3 Residential Satisfaction

While 63.5% of total households in Korea reside in apartments [25], challenges exist in building sustainable communities due to high density, architectural uniformity, and social disconnection among residents. In this context, residential satisfaction has emerged as an important research topic [26]. Residential satisfaction serves as a subjective evaluation of physical residential environment and resident interaction, functioning as a platform for physical, psychological, and social well-being beyond mere living space [27]. It is influenced by various factors including economic efficiency, convenience, environmental quality, social connectivity, and safety, and is evaluated based on the difference between individual subjective expectations and actual experiences [28]. Recently, it has expanded into socio-psychological dimensions, presenting a multidimensional approach integrating objective elements and subjective perceptions [29]. Recent studies emphasize the interrelationships between social capital, sense of community, and residential satisfaction, suggesting that residential satisfaction mediates between environmental attributes and behavioral outcomes [30]. Residential satisfaction is a multifaceted concept significantly affecting individual and family well-being, closely related to external social relationships and sense of community, representing residents' overall evaluation of residential units and surrounding environment through fulfillment of housing needs and expectations.

2.4 Mindfulness

Mindfulness has emerged as an important psychological concept for addressing stress and mental fatigue in modern society. Originating from Buddhist practice, mindfulness has evolved beyond religious context into conscious awareness that pays non-judgmental attention to present moment physical sensations, perceptions, thoughts, and emotions [31]. The Mindfulness-Based Stress Reduction (MBSR) program [32] developed by Professor Jon Kabat-Zinn has clinically applied and popularized mindfulness, being widely utilized as an evidence-based program helping people experiencing various difficulties including stress, anxiety, depression, and pain. Since the 1980s, increasing research has proven its therapeutic effects, with meta-analyses confirming that mindfulness practice effectively reduces depression, anxiety, stress, and pain, contributing to overall quality of life improvement through enhanced cognitive flexibility and emotion regulation [33]. Despite human cognition's subjective characteristics, mindfulness meditation helps develop objective and effective coping abilities for experiences by focusing on the present moment rather than worrying about past or future [34]. Thus, mindfulness plays an important role in enhancing psychological well-being and increasing adaptability to life's challenges.

2.5 Quality of Life (QoL)

Quality of Life (QoL) is a multidimensional concept encompassing both objective living conditions and subjective individual perceptions [35]. Beginning with social indicator research in the 1960s, it manifests diversely according to socioeconomic development, individual characteristics, living environment, and cultural values [36].

WHO defines it as perception of one's life position including physical, psychological, and social dimensions, emphasizing an approach integrating objective indicators and subjective measurements. Recent studies report that community interaction and residential satisfaction play important roles in quality of life, relating to community quality of life as subjective evaluation of residential environment [37]. Environmental perception significantly impacts quality of life, and quality of life research has developed toward integrating objective conditions and subjective satisfaction. Indicators like OECD's Better Life Index are being utilized in policy decisions, and quality of life is an important concept reflecting individual and family well-being [38].

3 Research Method

3.1 Research Method and Hypothesis Setting

This investigation examines how social capital influences quality of life among apartment residents, focusing on mediating effects of community sense and residential satisfaction, while considering mindfulness as a moderating factor between community sense and life quality. Contemporary research demonstrates social capital's significant role in community life satisfaction and health [21]. Studies have investigated various pathways connecting social capital to residents' life quality [7], with particular attention to mediating emotional factors [9]. Evidence suggests social capital's connection to community engagement enhances life quality [11]. This research investigates the distinct functions of community sense [23], residential contentment [28], and mindfulness [31]. The research framework appears in Fig. 1.

The hypothesis development builds on comprehensive literature analysis exploring relationships between social capital, community dynamics, and well-being. Integrating perspectives from Putnam's [11] social capital research, Chun's [26] apartment community studies, Bishop et al.'s [31] mindfulness investigations, and WHOQoL Group's [37] quality metrics, alongside recent Korean apartment community research, we propose these hypotheses.

H1: Social capital positively influences community sense.

H2: Social capital enhances residential satisfaction.

H3: Community sense contributes to residential satisfaction.

H4: Community sense improves life quality.

H5: Residential satisfaction elevates life quality.

H6: Community sense mediates between social capital and residential satisfaction.

H7: Community sense mediates social capital's effect on life quality.

H8: Community sense and residential satisfaction jointly mediate between social capital and life quality.

H9: Mindfulness moderates community sense's impact on life quality.

These hypotheses aim to illuminate how residential and social elements interact to affect life quality, seeking practical enhancement strategies.

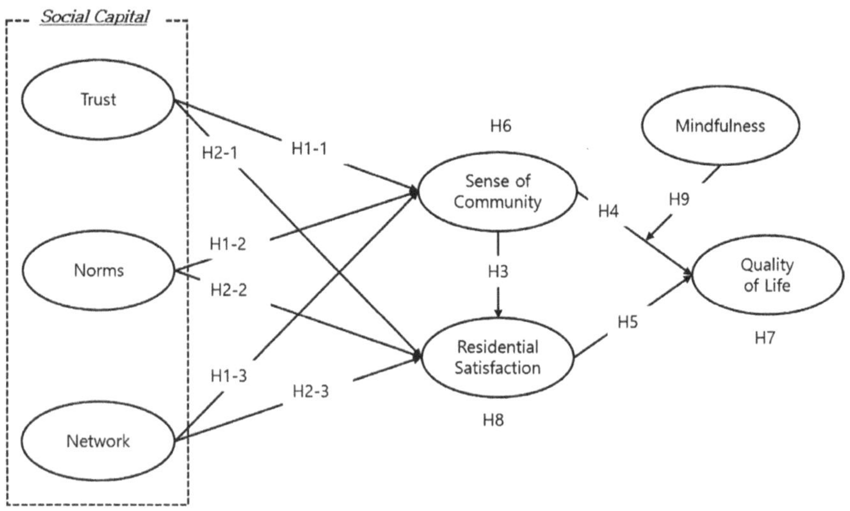

Fig. 1 Research model

3.2 Data Collection and Analysis Method

This research surveyed 378 apartment residents across Seoul, Gyeonggi, and Incheon from September 10, 2023, to November 9, 2023, utilizing face-to-face interviews, field visits, and digital surveys. Analysis employed SPSS 24.0 and AMOS 26.0, incorporating measurement validation through confirmatory factor analysis and hypothesis testing via structural equation modeling.

3.3 Measurement Tools

To test the hypotheses of the study, the operational definitions of various variables and measurement tools were organized based on prior research and literature review, and the operational definitions of the variables are presented in Table 1.

4 Research Results

4.1 Measurement Model Analysis

To analyze the fitness, reliability, and validity of the measurement model, we presented the measurement model as shown in Table 2 and conducted Confirmatory Factor Analysis (CFA). The goodness of fit test results for the measurement

Table 1 Operational definitions of variables

Variable		Operational definition	
SC	Trust	The expectation that members of the community will engage in regular, honest, and cooperative behavior based on universal norms	Putnam [11], Fukuyama [2], Lin [7]
	Norms		
	Network		
COM		The awareness and feelings of being a member of a community, formed through close interactions with neighbors and interdependence between individuals and the community	Sarason [22], McMillan and Chavis [23], Cheon [26]
RS		The degree of satisfaction an individual feels about their residence, based on the realization of expectations, desires, or aspirations	Lee [19], Jung [20]
Mindfulness		The intentional focus on the present moment's experiences, observing the body and mind with a non-judgmental attitude	Neff [5], Kabat-Zinn [32]
Quality of life		A concept encompassing all elements that make life valuable, including social welfare, life satisfaction, good living, and happiness	Campbell [35], Ventegodt et al. [38], WHO [37]

Note = SC: Social Capital, COM: Sense of community, RS: Residential Satisfaction Research Results

model showed $\chi2 = 576.689$, $\chi2/df = 2.074$, which is below the recommended level of 3, and RMSEA = 0.053, which is below the recommended level of 0.8. Additionally, AGFI = 0.869, which is above the recommended level of 0.8, and CFI = 0.953, TLI = 0.945, which are above the recommended level of 0.9, indicating that the fitness of the measurement model presented in this study was secured [39].

Table 2 Measurement model fit verification results

Category	χ^2	χ^2/df	RMSEA	AGFI	CFI	TLI
Model	576.689	2.074	0.053	0.869	0.953	0.945
Criteria	$p \geq 0.05$	$1.0 \leq Q \leq 3.0$	≤ 0.08	≥ 0.80	≥ 0.90	≥ 0.90

Table 3 Discriminant validity verification results

	1	2	3	4	5	6	7
1. Trust	**0.794**						
2. Norm	0.506***	**0.828**					
3. Network	0.517***	0.686***	**0.770**				
4. COM	0.485***	0.690***	0.595***	**0.749**			
5. RS	0.486***	0.621***	0.583***	0.697***	**0.738**		
6. MIND	0.702***	0.609***	0.578***	0.530***	0.500***	**0.834**	
7. QOL	0.789***	0.573***	0.560***	0.565***	0.543***	0.593***	**0.811**
AVE	0.629	0.686	0.592	0.560	0.605	0.706	0.711
CR	0.836	0.865	0.812	0.849	0.870	0.899	0.904

*** $p < 0.001$ Note = COM: Sense of community, RS: Residential Satisfaction, MIND: Mindfulness, QOL: Quality of Life. * The values in the shaded diagonal cells represent the square root of AVE, while the remaining values are correlation coefficients

4.2 *Discriminant Validity*

The criterion for verifying discriminant validity is based on the Average Variance Extracted (AVE), and it is established that the AVE value must be greater than the square of the correlation coefficient between two latent variables. Additionally, the AVE values for each variable were above 0.5, confirming discriminant validity, and the CR values were also above 0.7, ensuring reliability. The results of the discriminant validity verification for this study are presented in Table 3.

4.3 *Structural Model Analysis and Hypothesis Testing*

The second stage in the verification of the structural equation model (SEM) involves assessing the fit of the structural model. The structural model's fitness must precede the verification of causal relationships between latent variables. The indices selected to verify the structural model's fitness in this study are identical to those used for measurement model fitness verification. The structural model's goodness of fit test results showed $\chi^2 = 602.216$ ($p < 0.001$), $\chi^2/df = 2.762$, which is below the recommended level of 3, and RMSEA = 0.068, which is below the recommended level of 0.8. Additionally, AGFI = 0.857, which is above the recommended level of 0.8, and CFI = 0.927, TLI = 0.916, which are above the recommended level of 0.9, indicating that the measurement model's fitness proposed in this study was secured [39].

This study conducted hypothesis testing to analyze the structural relationships among social capital, sense of community, residential satisfaction, and quality of life through path coefficient analysis of the structural model, as shown in Table 4. The verification results are as follows.

Table 4 Path coefficient analysis results of the structural model overall sample (N = 387)

Hypothesis	Path			Path coefficient	Standard error	C.R	*p*	Acceptance
H1	Social capital → Sense of community							
H1-1	Trust	→	COM	0.184	0.055	3.194	0.001	ACC
H1-2	Norm			0.486	0.066	6.361	0.000	
H1-3	Network			0.183	0.073	2.474	0.013	
H2	Social capital → Residential satisfaction							
H2-1	Trust	→	RS	0.118	0.056	2.087	0.037	ACC
H2-2	Norm			0.159	0.070	2.020	0.043	
H2-3	Network			0.163	0.074	2.240	0.025	
H3	COM	→	RS	0.425	0.079	5.558	0.000	ACC
H4	COM	→	QOL	0.404	0.087	5.284	0.000	
H5	RS	→		0.286	0.081	3.915	0.000	

Note = ACC: Accepted, COM: Sense of community, RS: Residential Satisfaction, MIND: Mindfulness, QOL: Quality of Life

4.4 Mediation Effect Analysis

This study further analyzed the indirect(mediating) effects of sense of community and residential satisfaction on the relationship between social capital (trust, norms, network) and quality of life using phantom variables to verify statistical significance. The results are shown in Table 5.

Table 5 Individual path mediation effect analysis results

Path						IE		Confidence interval			Acceptance
						ES	S.E	LLCI	ULCI	*p*	
H6	Trust	→	COM	→	Quality of life	0.094	0.092	0.012	0.276	0.014	ACC
	Norm					0.215	0.107	0.094	0.357	0.004	ACC
	NET					0.091	0.082	0.005	0.014	0.023	ACC
H7	Trust	→		→	RS	0.015	0.010	0.002	0.198	0.013	ACC
	Norm					0.035	0.015	0.011	0.071	0.073	REJ
	NET					0.015	0.009	0.001	0.036	0.042	ACC
H8	Trust	→	RS	→	Quality of life	0.026	0.047	-0.004	0.120	0.159	REJ
	Norm					0.034	0.031	-0.004	0.110	0.133	REJ
	NET					0.047	0.032	-0.002	0.119	0.072	REJ
H9	COM	→		→		0.325	0.305	0.235	0.848	0.004	ACC

Note = IE: Indirect Effect, ES: Estimate, COM: Sense of community, RS: Residential Satisfaction, AC: Accepted, REJ: Reected, NET: Network

4.5 *Moderating Effect Analysis*

This study examined the moderating effect of mindfulness on the relationship between sense of community and quality of life using SPSS 24.0 and PROCESS MACRO. Analysis showed that both sense of community ($\beta = 0.346$, $p < 0.001$) and mindfulness ($\beta = 0.439$, $p < 0.001$) positively affected quality of life, with their interaction term showing significant effect ($\beta = 0.091$, $p < 0.05$).

The model was significant ($F = 3.989$, $p < 0.001$, $R^2 = 0.361$), explaining 36.1% of life quality variance. The R^2 change from interaction was 0.007 ($F = 3.989$, $p < 0.05$), confirming mindfulness's moderating role.

These findings indicate that higher mindfulness levels strengthen the positive relationship between community sense and life quality, supporting the hypothesis that mindfulness enhances community engagement's benefits on residents' quality of life [40]. The results are shown in Table 6.

5 Research Results

This study empirically analyzed the relationships between social capital, sense of community, residential satisfaction, and mindfulness to explore in depth the factors affecting the quality of life of apartment residents in Korea. The main findings and implications of this study, which aimed to contribute to understanding the quality of life of apartment residents who constitute more than half of the total population [25], are as follows.

Table 6 Moder results of the moderating effect of mindfulness between sense of community and quality of life

Variable	Coeff	SE	t	LLCI	ULCI	*p*
Independent variable (sense of community)	0.346	0.051	6.747	0.245	0.447	0.000^{***}
Moderating variable (mindfulness)	0.439	0.053	8.314	0.335	0.543	000^{**}
Independent * moderating	0.091	0.046	1.997	0.001	0.181	0.047^{*}
F(p)	$3.989(0.000^{***})$					
R^2	0.361					
ΔR^2	$0.007(F = 3.989, p = 0.047^{*})$					

*** $p < 0.001$, * $p < 0.05$

5.1 Research Findings

The structural equation model analysis of 387 apartment residents revealed complex relationships between social capital (trust, norms, networks) and quality of life [40]. The main research findings are as follows.

First, the sub-factors of social capital (trust, norms, networks) had a significant positive effect on sense of community and residential satisfaction [7, 12]. Second, sense of community had a significant positive effect on residential satisfaction, which was also found to have a positive impact on quality of life [23]. Third, residential satisfaction also had a significant positive effect on quality of life [35]. Fourth, social capital was found to influence quality of life through the sequential mediation of sense of community and residential satisfaction [19]. Fifth, mindfulness showed a significant moderating effect in the relationship between sense of community and quality of life [32], suggesting that the positive impact of sense of community on quality of life may vary depending on the level of mindfulness.

5.2 Academic Implications

This study supported and extended existing theories [12, 18] by empirically analyzing the relationship between social capital and quality of life among apartment residents. In particular, it deepened academic discussion by identifying sense of community and residential satisfaction as important mediating variables between social capital and quality of life [26], and revealing that mindfulness moderates the relationship between sense of community and quality of life [33, 35]. By analyzing variables as an integrated model, it contributed to understanding the pathways affecting quality of life in apartment residential environments and established a foundation for future research.

5.3 Practical Implications

This study presents practical measures to improve the quality of life of apartment residents [6, 11]. First, efforts are needed to enhance social capital and activate communities through resident meetings, cultural events, and online platforms [13]. Second, sense of community should be increased through participation in joint problem-solving to improve residential satisfaction and quality of life [15]. Third, the introduction of mindfulness programs and emotional support through community activities can support residents' psychological stability and quality of life improvement [9]. An integrated approach to strengthening social capital, sense of community, residential satisfaction, and mindfulness elements will be effective in improving quality of life.

5.4 *Limitations and Suggestions for Future Research*

This study confirmed the relationships and implications between social capital, sense of community, residential satisfaction, mindfulness, and quality of life through empirical analysis of apartment residents [19]. However, there are limitations in identifying causal relationships due to the cross-sectional research design targeting specific regions [16]. Future research needs to explore dynamic relationships and causality between variables through longitudinal/experimental studies including various regions and housing types [21], and consider other potential factors affecting quality of life [18].

References

1. Wang D, Wang F (2016) Contributions of the usage and affective experience of the residential environment to residential satisfaction. Hous Stud 31(1):42–60
2. Fukuyama F (1995) Social capital and the global economy. Foreign Aff 74(5):89–103
3. OECD (2001) The well-being of nations: the role of human and social capital, education and skills. OECD Publishing
4. Hwang YH, Park JW (2023) A study on the effect of apartment community servicescape on residential satisfaction and quality of life: focusing on newlywed hope town. J Korean Soc Qual Manag 51(3):329–346
5. Neff KD (2003) The development and validation of a scale to measure self-compassion. Self Identity 2(3):223–250
6. World Bank (2013) Inclusion matters: the foundation for shared prosperity. World Bank Publications
7. Lin N (2002) Social capital: a theory of social structure and action. Cambridge University Press
8. Ostrom E, Ahn TK (2003) A social science perspective on social capital: social capital and collective action. Revista Mexicana De Sociologia 65(1):155–233
9. Coleman JS (1990) Commentary: Social institutions and social theory. Am Sociol Rev 55(3):333–339
10. Bourdieu P (1986) The forms of capital. In: Richardson JG (ed) Handbook of theory and research for the sociology of education. Greenwood Press, pp 241–258
11. Putnam RD (1993) The prosperous community: Social capital and public life. Am Prospect 4(13):35–42
12. Samsung Economic Research Institute (2009) Social conflicts and economic costs in Korea (CEO Information No. 710). SERI
13. Gambetta D (ed) (1988) Trust: making and breaking cooperative relations. Blackwell
14. Douglas G, Pearce J, Woodward H (2007) A failure of trust: resolving property disputes on cohabitation breakdown. University of Bristol
15. Eisenbeiss SA, Maak T, Pless NM (2014) Leader mindfulness and ethical decision making. In: Neider LL, Schriesheim CA (eds) Advances in authentic and ethical leadership. Information Age Publishing, pp 191–208
16. Lee JY (2018) The effect of participation perception of elderly job and social activity support project participants on welfare consciousness: verification of mediating effect of social capital [Doctoral dissertation]. Seoul Hanyang University
17. Coleman JS (1988) Social capital in the creation of human capital. Am J Sociol 94:S95–S120
18. Granville DJ, Cassidy BA, Ruehlmann DO, Choy JC, Brenner C, Kroemer G, van Breemen C, Margaron P, Hunt DW, McManus BM (2001) Mitochondrial release of apoptosis-inducing factor and cytochrome c during smooth muscle cell apoptosis. Am J Pathol 159(1):305–311

19. Kim GH, Kim HJ (2016) A study on the effect of elderly lifelong education participation characteristics on social capital. Korean J Soc Welf Educ 33:29–57
20. Ell K (1984) Social networks, social support, and health status: a review. Soc Serv Rev 58(1):133–149
21. Szreter S, Woolcock M (2004) Health by association? Social capital, social theory, and the political economy of public health. Int J Epidemiol 33(4):650–667
22. Sarason SB (1974) The psychological sense of community: prospects for a community psychology. Jossey-Bass
23. McMillan DW, Chavis DM (1986) Sense of community: a definition and theory. J Community Psychol 14(1):6–23
24. Nisbet RA (1993) The sociological tradition. Transaction Publishers
25. Statistics Korea (2021) Metropolitan area apartment ratio. Statistics Korea
26. Cheon HS (2004) Social capital in metropolitan apartment complexes. Korean J Sociol 38(4):215–247
27. Kang ET, Jung HM (2015) The effect of residential characteristics on residential satisfaction of elderly households. J Residlal Environ Inst Korea 13(1):109–121
28. Galster GC, Hesser GW (1981) Residential satisfaction: compositional and contextual correlates. Environ Behav 13(6):735–758
29. Weidemann S, Anderson JR, Butterfield DI, O'Donnell PM (1982) Residents' perceptions of satisfaction and safety: a basis for change in multifamily housing. Environ Behav 14(6):695–724
30. Marans RW, Spreckelmeyer KF (1981) Evaluating built environments: a behavioral approach. University of Michigan Institute for social research and the architectural research laboratory
31. Bishop SR, Lau M, Shapiro S, Carlson L, Anderson ND, Carmody J, Segal ZV, Abbey S, Speca M, Velting D (2004) Mindfulness: a proposed operational definition. Clin Psychol Sci Pract 11(3):230–241
32. Didonna F, Kabat-Zinn J (2009) Clinical handbook of mindfulness. Springer
33. Goldberg SB, Tucker RP, Greene PA, Davidson RJ, Wampold BE, Kearney DJ, Simpson TL (2018) Mindfulness-based interventions for psychiatric disorders: a systematic review and meta-analysis. Clin Psychol Rev 59:52–60
34. Siegel RD, Germer CK, Olendzki A (2009) Mindfulness: what is it? Where did it come from? In: Didonna F (ed) Clinical handbook of mindfulness. Springer, pp 17–35
35. Campbell A (1981) The sense of well-being in America: recent patterns and trends. McGraw-Hill
36. Padilla GV, Grant MM (1985) Quality of life as a cancer nursing outcome variable. Adv Nurs Sci 8(1):45–60
37. WHOQoL Group (1993) Study protocol for the world health organization project to develop a quality of life assessment instrument (WHOQOL). Qual Life Res 2(2):153–159
38. Ventegodt S, Andersen NJ, Merrick J (2003) Quality of life philosophy I: quality of life, happiness, and meaning in life. Sci World J 3:1164–1175
39. Bae BR (2021) Structural equation modeling with Amos 27. Chungram Publishing, Seoul
40. Hwang YH (2024) A study on the impact of social capital of apartment residents on quality of life: focusing on mindfulness [Doctoral dissertation]. Soongsil University

The manufacturer's authorised representative in the EU is Springer Nature Customer Service Centre GmbH, Europaplatz 3, 69115 Heidelberg, Germany. If you have any concerns regarding our products, please contact ProductSafety@springernature.com

Printed and bound by CPI Group (UK) Ltd, Croydon, CR0 4YY
07/10/2026
02229367-0001